W9-BUQ-927

The Rise of Gold in the 21st Century

Edge Financial Company, Inc.
9730 SW Cascade Blvd., #200
Tigard, OR 97223-4324

also by James DiGeorgia

The New Bull Market in Gold:
$1,000 Gold and the Many Ways to Profit from It

The Rise of Gold in the 21st Century

Safety and Profits in an Age of War, Terrorism, Oil Shocks, Inflation, and the Decline of the Dollar

by James DiGeorgia and Don Mahoney

21st Century Investor Publishing, Inc.

Copyright © 2005 21st Century Investor Publishing, Inc.
All rights reserved.

This book, or parts thereof, may not be reproduced in any form without permission from the publisher; exceptions are made for brief excerpts used in published reviews.

Published by 21st Century Investor Publishing, Inc.
925 South Federal Highway Suite 500 Boca Raton FL 33432

ISBN 0-9718048-8-5

Library of Congress Control Number: 2004098139

Printed in the United States of America.

09 08 07 06 05 5 4 3 2 1

This publication is designed to provide accurate and authoritative information in regard to the subject matter covered. It is provided with the understanding that the publisher is not engaged in rendering legal, accounting, or other professional services. If legal advice is required, the services of a competent professional person should be sought.—From a Declaration of Principles, jointly adopted by a Committee of the American Bar Association and the Committee of Publishers and Associations.

Although the authors and 21st Century Investor Publishing, Inc. believe the information, data, and contents presented are accurate, they neither represent nor guarantee the accuracy and completeness nor assume any liability. It should not be assumed that the methods, techniques, or indicators presented in this book will be profitable or that they will not result in losses. Trading involves the risk of loss, as well as the potential for profit. Past performance is not a guarantee of future results.

Table of Contents

List of Figures vi
Preface vii
Introduction ix

Part I: Five Thousand Years of Value

1: Timeless Lure, Ancient Lore, Early Money 3
2: Gold by Conquest 13
3: Nazi Lust for Gold 35
4: Gold and Economics in the 19th and 20th Centuries 47
5: The California Gold Rush 59

Part II: Where Is Gold Found, How Is It Obtained, and How Much Is Left?

6: Modern Gold Production 71
7: How Much Gold Is Left to Be Found? 85

Part III: Modern Economics and Gold: The Pressure Mounts

8: Central Banks and Gold 95
9: Gold and the IMF 109
10: Current Economic Trends and Gold 117

Part IV: Investing in Gold

11: Direct Investments in Gold 141
12: Indirect Investments in Gold: Gold Stocks and Mutual Funds 157

Part V: Safe Haven from the Gathering Storm

13: The Rise of Gold in the 21st Century 173

For More Information 179
About the Authors 181
References 183
Endnotes 188
Index 191

List of Figures

Gold Production Since 1900 [6.1] 75
World Official Gold Holdings (Sep. '03) [8.1] 104
Largest Official Gold Holdings (Sep. '03) [8.2] 105
World's Proven Oil Reserves [10.1] 118
History of Oil Prices [10.2] 119
U.S. Percentage Imports for Top Five Suppliers [10.3] 120
1987 Stock Market Crash [10.4] 123
CBOE 30 Year Treasury Note Yield Index [10.5] 124
1994 Impact of Orange County Bankruptcy on S&P 500 [10.6] 125
Russian Default and the Long Term Capital Management Crisis [10.7] 125
Federal Debt [10.8] 126
Federal Budget [10.9] 127
Federal Receipts and Outlays [10.10] 128
Total Consumer, Corporate, Government Debt as Percent of GDP [10.11] 130
Sources of U.S. Debt [10.12] 131
Easy Money [10.13] 132
Trend of Inflation in the U.S. [10.14] 132
Existing Home Sales and Prices [10.15] 133
CBOE 30 Year Treasury Note Yield Index [10.16] 134
Price of Gold After Bottom in 2001 [10.17] 135
Dollar Decline Since 2002 [10.18] 136
Long-Term Chart of Gold [11.1] 142
Gold Mutual Funds [12.1] 163-4
Central Fund of Canada Price Performance vs. Gold [12.2] 164
Gold Stocks [12.3] 167-8
Long-Term Gold Prices [13.1] 176

Preface

This book is meant for those who have an interest in gold as a store of value and a safety net in uncertain economic and political times. It was written to satisfy and further educate those already convinced of the value of gold, and to educate the novice investor or collector.

It was also written from the beliefs and passions of the authors for the subject: gold.

The authors, James DiGeorgia and Don Mahoney, wish to thank a number of contributors to this book. Dan Hassey, Heather Robson, Spiros Psarris, Gordon Casey, Zara Altair and Curtis Tuck all provided solid research and contributed whole chapters. Their passion for the subject comes through in their writings.

It is also to honor their writing that only a minimal effort has been made to edit this book into one "voice." All the contributors—besides their contribution of subject matter—added their particular style to the mix. As a result, different sections contain different tones and style. We believe the styles and voices add to the final result, not detract.

You can read this book from front to back, back to front, or even simply Part IV, or Part III and IV, or any other way you like. We believe it will work any way you read it.

Once again, thanks again to the contributors. In all cases, their work was first rate.

Introduction

"Water is best. But gold shines like fire blazing in the night, supreme of lordly wealth."
—Pindar (c. 518-438 B.C.) Olympian Odes I

"Gold is a barbarous relic."
—economist John Maynard Keynes

$7.59 million
The price paid in 2003 for a gold U.S. 1933 $20 Saint Gaudens Double Eagle coin, containing less than one ounce of gold. It is currently insured for $10 million.

68%
The increase in the price of gold between July 20, 1999 and January 13, 2004.

34.4%
How much the U.S. Dollar Index fell between 1986 and early 2004.

* * * * * * *

Gold.

It has been coveted by man since the dawn of history.

Perhaps Keynes was right, and Pindar wrong. Even in the 1990's, as the stock market soared, pundits proclaimed, "Gold is dead."

But was it? Or was it only sleeping?

If it was dead, how could one ounce fetch $7.5 million in 2003? Granted, it was a rare coin, but its value had much to do with it being a rare *gold* coin.

And if gold is a barbarous relic, or even dead, how could it rise 68% from July 20, 1999 to January 13, 2004, even while the S&P lost 50% from March 24, 2000 to October 10, 2002?

(If you owned $100,000 in S&P stocks during that slide, you lost $50,000. If you owned $100,000 in gold bullion during that bull run in gold, you gained $68,000. A difference of $118,000.)

Gold was used to make a coffin for King Tutankhamen. It has adorned necks, filled teeth, caused wars, and triggered the largest migration of people in the history of the world. It's been the source of countless schemes and dreams throughout history. And it is still stored as wealth by millions of people, worn on the bodies of billions of people, and is the target of thieves, burglars, and pilferers worldwide.

The Yellow Metal

Gold is a soft, dense, bright yellow metallic element. Its symbol is Au, from the Latin *aurum,* meaning "gold."

Gold can be found as large nuggets, or fine particles. The largest single mass of gold is called the Holtermann Nugget, found in Australia in 1872, weighing almost 290 kilograms. The rarest form of gold is a true nugget, and the largest known true nugget (not just a mass), is called the Welcome Stranger, weighing 2,284 troy oz. It was found accidentally, just below the surface of the ground, in Victoria, Australia, in 1869. Between them, their value is some $2,700,000 at recent prices.

Gold is also present in seawater, but in such miniscule quantities that the cost of recovery would be far greater than the value of the gold that could be recovered. Its presence is between 5 to 250 parts by weight to 100 million parts of water. There may be more than 9 billion metric tons in the seas, but you would have to process oceans to recover it.

But why has gold been the focus of human admiration for thousands of years? Why is gold so valuable?

Five thousand years of value

Gold's beauty has been recognized for thousands of years. But there's more to its appeal than that.

Gold is a tremendously useful metal—used in everything from medicine to electronics. But there's more to its appeal than that.

Gold has served as a medium of exchange—money—for countless civilizations. But there's more to its appeal than that.

Gold's value comes from *all* of these things combined. Gold's unique qualities make it ideally suited for each of these roles, and as a result it's been coveted and treasured for millennia.

There's no great mystery about this. Different materials are best suited for different things. We make books from paper, not from aluminum or bricks—because paper works best. In similar fashion, we make airplanes from aluminum, and houses from bricks.

When it comes to wealth, gold happens to be the material best-suited for storing and exchanging it. Mankind has tried many materials for these purposes: salt, land, cattle, food, even sticks and stones. But gold is best.

Gold has unique properties. It's the most malleable of metals: an ounce can be hammered into a translucent, one hundred square-foot sheet, a mere hundred-thousandth of an inch thick. It's incredibly ductile: an ounce can be drawn into a fine wire 62 miles long.

Gold has excellent conductivity, both thermal and electrical. It's highly reflective. And it's resistant to most forms of corrosion—even most acids can't hurt it.

More than two thousand years ago, Aristotle wrote about the "perfect money." He realized that the best money would have these characteristics:

- **Useful**. The money must have inherent value, so a piece of it would be worth something beyond its role as money. Sticks, green paper, and other forms of money fail here. But gold has been tremendously useful throughout history, and remains so today. It plays a vital role in electronics, medicine, dentistry, jewelry, photography, and many other modern industries.
- **Durable**. The money must be impervious to rot, decay, mildew, tarnish, and rust. Food and other forms of money fail this test. You don't want your wealth wiped out by a flood, a fire, or a swarm of insects. But gold is impervious to almost anything.
- **Fungible**. One ounce of pure gold is the exact same as any other ounce of pure gold, so it's easy to trade. Not so for artwork, jewelry, or land; every piece is unique and must be individually appraised.
- **Divisible**. Gold can divided into tiny quantities. But you can't make change with a piece of land, or a diamond ring.
- **Convenient**. Gold allows you to carry a lifetime of wealth in your pockets. Not so for copper, cattle, or a silo full of grain.

Although this didn't occur to Aristotle, there's another crucial reason why gold is the perfect money. Unlike paper currency, governments can't print up endless amounts of gold!

So why isn't gold still used as money? That's discussed elsewhere in this book. But in a sense, it really doesn't matter—all of gold's wonderful benefits are still available to us today as an investment, whether or not a government has stamped a dollar figure on the coin. All the historical reasons to store your wealth in gold are still valid—there are even some new ones today, as you'll see.

Why This Book?

There are plenty of good books on gold. *The Power of Gold* by Peter L. Bernstein. *How To Profit From The Coming Boom In Gold*, by Jeffrey A. Nichols. *Mining Explained: A Layman's Guide*, edited by James Whyte and Thomas Brockelbank. *The New World of Gold*, by Timothy Green.

So why this one?

The authors believe that the importance of gold has grown exponentially just in the last few years. There are growing economic problems facing the United States that will have a major impact on the markets and gold, for example:

- Growing dependence on foreign oil, threatening an oil shock
- The devolution of banking and the Federal Reserve Bank
- Proliferation of derivatives
- U.S. budget deficits
- Ballooning consumer, corporate, and government debt
- Rising inflation
- The falling dollar

We'll cover all of these in detail in this book.

However, this book is not intended to be a complete study of everything there is to know about gold. Our purpose is specific.

A Call to Action

We believe that it is more important than ever to own gold.

Not to sink every penny you have into the yellow metal. But to put somewhere between 10% and 25% of your assets and/or savings into gold. We believe this is the wisest possible investment in today's environment.

This is not an academic exercise for the authors. We practice what we preach, and we own gold. For all the reasons we espouse in the pages of this book.

And so we have tried to briefly communicate the importance of gold throughout history, and demonstrate its world-wide appeal and time-tested value as a store of wealth. We have tried to convey the passion of people throughout history for gold, the lengths they will go to acquire gold, the basics of gold and economics, and the basics of buying and holding gold for financial protection. Above all, to show why gold is such a compelling investment today.

Read.

Hopefully, enjoy.

Think.

If you agree, and haven't already done so: act!

Part I

Five Thousand Years of Value

CHAPTER 1

Timeless Lure, Ancient Lore, Early Money

From the beginning of time, the glint of gold has allured men to power and wealth, and motivated them to sacred and profane acts. Gold coinage is the origin of our monetary system, and today it flies across international borders electronically. Gold has inspired both poetry and murder...it's been the staple of peaceful trading, and the object of war... it's decorated the flesh of the living, and the tombs of the dead. Gold has been formed into necklaces, bracelets, buttons, and crowns...cast into bricks, and beaten into fine decorative leaf...minted into coins, and woven into cloth.

For thousands of years, gold has influenced the course of human history. The first recorded uses of gold come from Egypt. It was here that mining first developed as an industry, and slaves were first used to dig it from the ground. Goldsmithing became an art.

The use of gold in Egypt was a royal prerogative, unavailable to anyone but the pharaohs. Egypt's gold mining industry was already old at the time of King Tut. The first recorded creation of gold "money" occurred in Egypt, when Pharaoh Narmer (the first pharaoh) minted gold bars to a standard fourteen-gram size, stamped with his name. (However, this unit was too valuable for ordinary commerce—in today's terms, approximately $180 each at a gold price of $400 an ounce.)

For many centuries, Egypt was the world's leading gold producer. Its gold came from two areas: the plateau between the Nile and the Red Sea, and the southern kingdom of Nubia. (*Nub* was the Egyptian word for gold.) Nubia remained a source of gold, even to the Western world, well into the sixteenth century.

Gold was vital to other ancient cultures as well. Gold is the first metal mentioned in the Bible. The land next to Eden is "Havilah, where there is gold." The book of Genesis says that Abraham was very rich—not only in cattle, but in silver and gold. In the Book of Exodus, the sacred vessels of the Jewish tabernacle are listed: among them are the gold-covered ark, made to hold the sacred commandments, and the seven-branched candlestick which contained a talent (a large denomination which equaled about 100 pounds) of pure gold.

But stores of wealth sit. Money moves. Before gold could be used as money as well as a store of wealth, people had to become sufficiently productive to have something to trade, travel had to become more routine, and measurement had to become accurate. When gold was only a store of wealth, payments from one party to another were infrequent. The process was cumbersome and time-consuming.

Gold and silver, as the two most valuable metals known, became media of exchange before the coin itself was invented. Pieces of these metals were passed back and forth in trade, with each merchant possessing a scale for weighing the metals. In Babylonia (circa 2000-1600 B.C.), bits of gold weighing 8.34 grams each were known as shekels. In appearance, they look like small pieces of yellow dough which might have been formed by rolling between fingers. (The Hebrews also had a gold shekel—theirs weighed about 16.4 grams.) But anyone who accepted a shekel gambled on its quality. The purity of the gold varied from shekel to shekel. Buyer and seller had to haggle over the worth of the shekel as well as the goods. In other words, both seller and buyer had to beware.

Even the legend of the Golden Fleece (circa 1400 B.C. or earlier) apparently has a connection with reality. You might remember the story of Jason and the Argonauts, who stole the Golden Fleece from Colchis (where it was guarded by a sleepless dragon). The people of Colchis on the coast of the Black Sea gathered gold from the rivers by dipping sheepskins into the water, collecting the particles of gold that clung to the tight crimp of the wool. This seems to have been the origin of the myth.

The first gold coins appeared in the seventh century B.C. in Lydia. They were called *staters*, and they were minted by a usurper to the throne, Gyges, who established his power through gold and conquering neighboring lands. Officially, his successor Croesus (560-546 B.C.) created the first bimetallic system. He ordered the minting of one coin of pure gold, another of silver, and fixed the ratio between them at 13.33 silver coins to 1 of gold.

The coins were crude, more oblong than circular, with the head of a lion facing a bull stamped on one side. That stamp meant that coins had the support of a powerful Mediterranean kingdom, with Croesus guaranteeing the weight and purity. Traders accepted these coins without question. Gold had become a monetary metal, with silver as its junior partner.

Gold made commerce much easier. A ship captain with only a bag of coins as his cargo could make a voyage to distant lands, and he could return with a shipload of merchandise. A producer with a surplus of goods could sell that surplus for gold or silver coins, and those coins would serve as a store of value long after the merchandise might have rotted.

Lydian kings realized the value of honesty in coinage, and they adhered to high purity standards when they minted their staters. So did the Persians with their darics, the Athenians with their owls, and Alexander the Great with his staters.

Darius the Great, the Persian ruler who built a kingdom from the Mediterranean to the Indus River Valley, carried coinage a step further from the staters of the Lydians. He had his own portrait stamped on the coins minted in his kingdom. At the time, a king was viewed like a god, a figure meriting reverence. A coin with a representation of Darius on it, usually shown as a hunting bowman, indicated the presence of Darius in spirit if not in fact. So, every time a citizen handled one of these coins, he also had a reminder that he had better behave, lest the mighty Darius punish him.

Under Darius' rule (522 to 486 B.C.), the coin economy became a force in everyday life. The bimetallic system of Croesus had only been established around 560 B.C. In less than half a century, the system had spread throughout Asia Minor and into the kingdom of Persia.

Coinage added a new dimension to taxation. Before, taxes had been imposed in kind, with the ruler's warehouses collecting and holding grains, wines, and other products. Now the Persians started accepting coins for tax settlements, as well as products. From that day on, governments have had their hands in the money of citizens.

The Persians had so much gold that they bribed enemies. It was easier to buy off their enemies than fight them. In 525 B.C., when Persia added Egypt to its empire, it found itself in control of most of the gold-producing regions of the ancient world. Only Carthaginian-controlled Spain lay outside its sphere. Gold flowed into Persian coffers from Egypt, Anatolia, India, Afghanistan, and central Asia. Trade developed on a scale unknown before.

Though Persians were masters of gold diplomacy, they were no match for Alexander the Great, either at monetary diplomacy or at war. In the fourth century B.C. he came close to cornering the gold market. He loved gold so much that when he went into Asia (334 B.C.), along with his forty thousand troops he had a skilled mining engineer. Every time a city fell to his army, Alexander seized the civic treasury.

Persians had coined money conservatively, holding surplus in bars in their treasuries. But Alexander reversed monetary policy. He ordered gold and silver bars to be minted into coins. Then he distributed the coins to court favorites, and also created a professional army of paid soldiers. As he built new cities, he poured money into the economy of the known world.

He minted money faster than goods could be produced to match. As a result, Alexander created the first known monetary inflation. The drachmas and staters he minted became the first worldwide currency. The coins circulated from the Nile to the Indus and from the Black Sea to the Indian Ocean. Alexander's coinage system prevailed for over 150 years, from India to most of the Greek and Egyptian areas to the west, until the Roman general Quinctius Flaminius defeated Philip V in 197 B.C. and brought closure to Macedonian hegemony.

With diminished Greek power after Alexander, Carthage in North Africa became the dominating western empire. Gold-poor Rome began to notice the wealth across the Mediterranean, especially the rich mines in Spain that were held by Carthage. Spain became a Roman province and Carthage was destroyed by the end of the Punic Wars (264-146 B.C.).

The Romans created a monetary heritage that in some ways we still follow today. In 390 B.C. the cackling geese that lived around the temple of Jupiter alerted the Romans to a surprise attack by the Gauls, who were invading Italy. The Romans were so grateful for this notice of impending danger that they constructed a shrine to their goddess of warning, whose name was Moneta; *moneta* in turn became the foundation of our words "money" and "mint."

The Romans also gave us the monetary denomination of the pound—*libra*—which is why the pound sterling is identified by the symbol £. In addition, the Latin word *denarius* came to stand for penny and was conventionally abbreviated as *d* in English usage. Finally, the term *solidus* meant that a coin was pure gold or silver, was worth one-twentieth of a pound of silver, and was equivalent to twelve *denari*. English money was built up from these ratios. A pound was equal to twenty shillings and a shilling was equal to twelve pence—a system that lasted from Norman times until the 1970s.

On the negative side, the uncontrolled greed of some Roman emperors brought coin debasement to new levels. They added copper, zinc or other base metals to gold and silver coins. (They were not the first: Dionysius of Syracuse [405-367 B.C.] had borrowed money from his citizens and could not pay them back. He ordered all coins in the city brought to him, under penalty of death. He restamped the coins so that each one-drachma coin now read two drachmas. After that, paying off his debts was easy.)

Nero was the first emperor to use debasement for his own greed. As paper money and bank credit had not yet been invented, debasement was the only available method to "create" enough purchasing power to satisfy his constantly expanding needs. The Roman emperors learned to make debasement a routine procedure. This created a continual inflation as prices rose to compensate for the currency's loss of value.

Continued inflation, weak currency, and weaker emperors all led to Rome's vulnerability. The Roman Empire decayed and eventually fell. When Rome was taken, Western Europe fell into the Dark Ages.

With Rome's decline, Constantinople became the new capital of the world, the empire of Byzantium rising from the ashes of Roman civilization. The Emperor Constantine established the city on the Bosporus in 325 A.D. In the same year, he reformed the imperial currency, creating a new gold piece, the solidus or bezant. This unit, which contained sixty-five grains of gold, became the key currency of Byzantium.

The bezant was used by merchants from China to Ethiopia and from the steppes of Russia to the walled cities of the Adriatic. It became the epitome of sound money, a currency that retained its worth for nearly eight hundred years. No other currency in all history can match that record.

For over one thousand years, the Byzantine empire based in Constantinople replaced Rome as the religious and political center of Europe. The Byzantine emperors were preoccupied with gold; although they were debased morally and politically, the golden bezant remained pure and acceptable. The bezant financed the empire's imports, armies, and alliances with other nations.

Because the uses of gold in the empire were so varied, the skilled goldsmiths of Constantinople were in demand throughout Europe. Indeed, they were the principal artists of the Dark Ages.

The emperors amassed hoards of gold coins and gold bars: useful in trade, vital in war. When the Lombards threatened not only Byzantium but also all of Europe, the empire bribed the Franks with fifty thousand gold *solidi* for their help. The Franks conquered the Lombards, and it was their kingdom (flush with all that gold) that led the way back to gold coinage in Western Europe.

After Constantinople fell to the Crusaders in 1204, the bezant began to lose its purity to debasement. Because of this, the coin also started losing its wide acceptability. The Muslim countries, continual enemies of the Byzantine Empire, also hoarded gold, used it for display and decoration, and even imitated the bezant, replacing the image of the emperor with quotations from the Koran. This imitation gave their coins immediate acceptance. Because of the trade with Europe these coins, 97 percent pure gold and minted in large quantity, gradually displaced the bezant as the major international currency.

The Muslims obtained gold from West Africa through trade. The area was so rich in the yellow metal it came to be known as the Gold Coast. Around the year 750, the Muslims had even tried to invade these territories and capture the source for themselves—which failed miserably. Thereafter, they contented themselves with trading. (Interestingly, the main instrument of trade for West African gold was salt. Salt was so precious to the gold diggers that in many transactions, an ounce of gold exchanged for an ounce of salt.)

In the middle of the 13th Century the rising Italian trading powers of Florence, Genoa, and Venice began issuing gold coins that gained much repute. By the middle of the 14th Century, citizens of Byzantium were paying taxes, not in bezants, but in Venetian gold ducats.

During the middle ages, in spite of vast fluctuations in the values of gold and silver as supplies flourished or dried up, monarchs evolved monetary systems that have developed into the current financial system. European nations began to shift from operating mines to accumulating already existing gold through trade, bribes, and conquest, and employing it to enhance power and wealth.

By this time, Western Europeans had had their own coinage for centuries. For example, in eighth century England, "moneyers" hammered silver into pennies which circulated all over Europe because of their purity. But gold coins were so valuable in the Middle Ages that they did not circulate much among common people. The gold coins were used in transactions by merchants and traders involved in foreign trade, by tax collectors, by the monarch's retinue, and by monarchs as a means of buying off enemies and ransoming friends and family members. They all made sure that gold coins remained at levels of purity of weight and composition.

The English, especially, kept a tradition of "sound" currency while their continental neighbors made do with irregular currencies. The monarch had a strong vested interest in maintaining the purity of coinage because coins were almost the sole method of effecting transactions and paying taxes and debts.

The merchant cities of Italy had also become influential. Genoa was one of the first cities to use economic means (rather than military) to enhance its power. The city came up with its own solid gold coin, the *genoin*, which circulated among merchants and the upper classes.

Florence soon followed with the *fiorino d'oro*, the florin. In 1284 the Venetians issued the ducat, which served as a standard of value throughout Europe and maintained its gold content until the Venetian Republic fell to Napoleon in 1797. All of these coins weighed 3.5 grams and were 24-carat gold.

The fourteenth century stands as a time of unrelenting famine, pestilence, social chaos and warfare. As the Black Death took up to a third of the population of Europe, possessions and monetary wealth remained behind. The survivors were far richer than they were before. Uncertain times brought on a lavish spending spree with no incentives for saving. Clerics as well as others had huge appetites for imported luxuries like fancy foods and decorative, if not ostentatious, clothing. Ironically, the supply of precious metals soon did not meet demand. Mints were inactive for lack of precious metals. Many nations created "sumptuary" laws, which forbade the use of gold as personal adornment in order to keep private citizens from using gold and keep the gold in royal pockets.

The wars and lack of gold resulted in huge ransom demands from one nation to another in order to build supplies of gold. A captured king would be released only after payment of large ransom sums. So gold went from one country to another as wars were won and lost, and key figures bought their freedom.

When the fifteenth century began, gold supply was far below the demand. Nations began to seek new sources of gold in other parts of the world. Commerce and industry revived. Italy fared best, and within Italy, Venice grew most of all. Territories remitted a million gold ducats a year to the Venetians. But the Venetians exported a ton of gold a year in trade, reducing the continental gold supply.

As gold and silver supplies dried up, local payments were often made through barter. But barter items, such as pepper, fluctuated wildly in value. A few bags arriving at a harbor could depress value instantly.

Because of the lack of coins, commodity values fell while gold prices rose. The purchasing power of gold doubled during the fifteenth century. This was one of the few periods in history when gold was spent instead of hoarded. The pressures from lack of gold sources opened up exploration for gold, and the wealth and power it would bring. (These explorations are covered in the next chapter.) The burning need for gold was the critical stimulant for the voyages into new and uncharted territories—the Americas.

By the end of the sixteenth century paper issued by private parties instead of coins issued by governments became the expedient for trade. While monarchs were struggling with wars and outflows of gold, merchants were evolving a new way to conduct transactions across vast distances.

During this century, prices rose. Inflation reared its nasty head. However, the citizens of Europe had no experience with inflation. Economists call this period the Price Revolution of the Sixteenth Century. The purchasing power of money and of labor incomes deteriorated. Huge fiscal deficits and exploding government indebtedness ensued. In Spain and France governments borrowed from the capital markets, which supplemented the privately negotiated debts piled up with bankers in Italy, Germany and Holland. Plus, the older form of finance still flourished—devaluating currency. Henry VIII's devaluation in England was so blatant it was awarded upper case titling—The Great Debasement.

While monarchs and laborers were suffering, merchants thrived in large centralized trade fairs. Moneychangers flourished. But merchants became so frustrated trying to make arrangements in almost 50 different kinds of gold coins that increasingly paper instruments took the place of payment in coin.

Promissory notes, "bills of exchange," and other paper instruments revolutionized trade in the sixteenth century. Debts were still denominated in gold, but less gold exchanged hands. Instead, the paper would be traded instead, and only net differences would be settled with coins. The concept of money was being transformed. The nature of gold's role in the system began to change. It retained its value but began to be represented by paper.

Over time, gold coins circulated less frequently, and gold bullion served only to settle up very large transactions or to cover unfavorable balances of trade between Europe and the Far East. One problem of coin stability was known as *clipping*. People came up with ingenious ways to cut off or "clip" small pieces from coins. They would then hoard the gold or silver and some even made counterfeit coins from their clippings.

Since the time of the Lydians and Greeks, coins had been carefully hammered out by hand one at a time, resulting in coins with smooth edges. So coin clipping was noticeable. However, clippers learned to file the edges smooth to disguise their theft. In France, toward the end of the sixteenth century, horse drawn machinery was used to roll, press and cut coins with markings or inscriptions on the edges. Besides revolutionizing production of coins, the edges of the new machine-made coins made clipping actions readily apparent to the naked eye. The revolutionary process took hold over much of Europe.

The horse-driven machine process facilitated the creation of large amounts of coins. Now, it was much easier for countries to issue large amounts of coins quickly.

In 1663, Charles II of England issued a warrant to create a new gold coin produced entirely by the mechanized methods. The "guinea" was stamped with an elephant to represent the gold that came from West Africa. Soon new issues of silver coins were manufactured by the same method.

Coinage did not stabilize, however. Silver coins fluctuated wildly in value, mainly toward the down side. Clipping continued with a vengeance. And the obstacle to stabilization through total recoinage was that neither the crown nor the people wanted to bear the cost of the difference between face value and true value of the clipped and worn coins.

Thirty years later clipped coins were forbidden for any type of payment. The immediate result was that commerce ground to a halt and people did not pay all their taxes.

In 1689, hostilities broke out between England and France. By 1697 the English king, William III, was greatly in debt. Taxation, personal loans and lotteries did not meet the overwhelming demand for money. The result was the establishment of the now famous Bank of England. Men of "quality" became shareholders in the Bank, which loaned the government £1.2 million at 8 percent. In return the institution was established as the first private company to do business as a limited-liability corporation, or joint stock company.

In 1695, speculation in the guinea finally forced the government to reissue coins, setting the standard for the guinea against the still undervalued silver shilling. The Great Recoinage restored the weight of English money to a standard equal to that before The Great Debasement 150 years earlier.

The guinea's consistent weight and fineness held true against silver coinage, which became greatly undervalued. Bankers held the guinea as reserves; tax collectors welcomed the solid guinea against bickering over the true value of silver coins. During the seventeenth century, in addition to the circulation of the government's guinea, private paper money began to substitute for coins in large transactions. "Bills of exchange" (private debt instruments) were often endorsed over from one holder to another and became a type of paper money. Also, people deposited their gold coins with a goldsmith for safekeeping and accepted in exchange a receipt for the gold that could be used as a means of payment. All of these paper instruments could be redeemed for gold at any time.

By the end of the seventeenth century England's relationship to gold had set in motion coinage, paper instruments, and a banking institution which set the stage for traditions which continue to the present. Paper instruments representing money were a precursor to today's electronic trading. The Bank of England eventually became a model for central banks in other countries (see Chapter Eight). And the guinea was a true direct descendant of Croesus' stater, Constantine's bezant, the ducat, the genoin, and the florin. It remained the basic gold coin of England for another one hundred years.

CHAPTER 2

Gold by Conquest

"Pirates!" screamed the lookout. "Pirates to starboard!"

Captain Tomás de Alzola's heart sank. He knew what would come next.

First, the chase. But his galleon, the Santa Ana, *was slow and overloaded with gold, silver, and other treasures—and the pirates would catch him eventually.*

Then, the demand to surrender. If he agreed, he and his men would become prisoners and held for ransom—if they survived at all. If he refused, then would come the whistle of incoming cannon shot...the screams of men, amid the explosions and showers of hardwood splinters...and eventually, the clanging of swords as his men fought hand-to-hand with the boarders.

Grimly, he gave his orders.

* * * * * * *

As the ultimate form of wealth, gold has inspired much conflict in human history.

Some conquerors have been successful in their quest for gold. In the last chapter, we saw how Alexander the Great built his empire, driven partially by a lust for gold.

At other times, would-be conquerors failed in their objectives. Marcus Crassus (115-53 B.C.), a Roman general, invaded Parthia to seize its golden treasure. His army of eleven legions was overwhelmed and Crassus fell. The victorious Parthian king ordered gold to be melted immediately, and the molten gold was poured down the throat of the Roman general. The last words Crassus heard were, "You have thirsted for gold, therefore, drink gold!"

Sometimes in history entire wars have raged specifically for gold. The broadest example of this must be the wars fought over gold in and around the New World discovered by Christopher Columbus (the Americas). At no other time in history has such a conflict occurred: waged both by governments and on individual initiative…involving many nations and several different people groups…battles fought over land and at sea…and spanning more than 200 years.

It all started in 1492…

* * * * * * *

> *"Gold is the most excellent of all commodities. Gold constitutes treasure, and he who possesses it has all he needs in this world, as also the means of purchasing souls from purgatory and restoring them to the enjoyment of Paradise."*
>
> *Christopher Columbus*

Schoolbooks tell us that in 1492, Columbus set sail to find a western passage to India. Europeans had been fascinated with the riches of the East since Marco Polo had returned from there two centuries earlier. However, trade for the spices, silks, and other wealth of India and Southeast Asia was mostly over land—so these goods passed through many hands along the way, which raised prices considerably. The Europeans—to eliminate these middlemen—were very interested in establishing sea routes directly to the East.

Many efforts were made to explore along the coast of Africa, to find a passage around the continent. In 1488 the Portuguese were the first (in the person of Bartolomeu Dias) to round the Cape of Good Hope, although nobody actually made it all the way to India until Portuguese navigator Vasco da Gama did it in 1497.

Columbus took a different route. Contrary to what many believe, he was not trying to prove that the Earth was round—that was widely known in his day. Instead, Columbus believed that the Earth was smaller than most of the cartographers of his day had calculated—therefore, he thought Asia must be within range for the ships of that time. As it turned out, the cartographers were right and he was wrong, but there were other continents within range instead—the Americas.

Columbus first landed on an island in the Caribbean (which he named San Salvador—probably today's Watling Island). He also explored Cuba and the island which became known as Hispaniola (today made up of Haiti and the Dominican Republic). Columbus mistakenly referred to these islands as the "Indies," as he believed he was close to India.

Although the real Indies held many allures—spices, silk, luxuriant rugs, etc.—the Europeans were most desirous of gold. King Ferdinand himself had instructed, "Get gold, humanely if possible, but at all hazards—get gold." Columbus needed no encouragement in this regard. "From October 12, 1492, to January 17, 1493, when Columbus headed back to Spain, his diary mentions gold more than 65 times. Indeed, his entry of October 13, 1492, the day after the first landing, reports, 'I was attentive and took trouble to ascertain if there was gold.'"[1] In writing to Ferdinand and Isabella about the native peoples he had found, Columbus said, "So your highness should resolve to make them Christians, for we believe that…you will achieve the conversion to our holy faith of a great number of peoples, with the acquisition of great lordships and riches and all their inhabitants for Spain. For without doubt there is in these lands a very great amount of gold."[2]

The Spanish wasted little time in conquering the Indies—an expedition was sent to settle Hispaniola in 1493, and others followed soon thereafter. Their motives were numerous: conquest, fortune, and the excitement of a new frontier. But, as one author comments, "Gold was of course the main attraction to the Spaniards. Gold they sought from the first, craving it, pursuing it obsessively."[3]

As new islands were discovered in the Caribbean, the first explorers were given vast estates on them. Thus, as more settlers arrived, the choicest locations were already taken, leading to constant pressure to explore further and conquer more territory. Eventually, the coastline of Central America was being probed by the Spanish as well.

Some of the native peoples there were hostile, and many Spaniards were killed in initial contacts. However, the Spanish persisted. Early on, a Spaniard by the name of Jeronimo de Aguilar was shipwrecked on the mainland and "went native," learning the Mayan language and marrying a Mayan wife. Years later, when the Spaniards found him, de Aguilar enabled the Spanish to communicate with the native populations. Soon the Spanish learned that there was a powerful civilization to the west, rich with gold and other wealth—the Aztec empire. But as the Spanish explored further, they were unprepared for what they would find…

****

"When Alvarado came to these villages he found that they had been deserted on that very day, and he saw in the cues [temples] *the bodies of men and boys who had been sacrificed, the walls and altars all splashed with blood, and the victims' hearts laid out before the idols. He also found the stones on which the sacrifices had been made, and the flint with which their breasts had been opened to tear out their hearts.*

"Alvarado told us that most of the bodies were without arms and legs, and that some Indians had told him that these had been carried off to be eaten. Our soldiers were greatly shocked at such cruelty. I will say no more about these sacrifices, since we found them in every town we came to…" [4]

The Aztecs were a proud civilization, on the one hand impressive for their achievements—while on the other, built on a brutal foundation of tyranny and oppression. The Spanish conquest of this mighty empire was driven by a strange mixture of motivations: a sincere desire to Christianize the people and free them from their barbaric, savage culture went hand-in-hand with a crude lust for their gold.

In February of 1519, Hernán Cortés set out to the mainland with an expedition of 11 ships, for the express purpose of conquering the Aztecs. To defeat this empire of millions of people, Cortés had 400 soldiers, 16 horsemen, and a few cannon. This seems either the height of arrogance, or foolishness—maybe both—except that nine months later, he was occupying the Aztec capital!

Cortés' expedition started out inauspiciously. Technically, he was in rebellion against his superiors—his commission from Diego Velázquez (the Governor of Cuba) to lead the expedition was apparently revoked before he left. This caused him problems later, as some of the men on the expedition were still loyal to Velázquez. His famous act of scuttling his own ships after arrival on the mainland was done to reduce the dissension among his men.

After landing on the coast, Cortés founded a city which he named Vera Cruz. He dedicated it to His Majesty Charles V, in an effort to place himself directly under Charles' authority—and to remove himself from that of Velázquez.

Cortés then spent the next several months exploring the surrounding area and making contact with the natives. He found that some of the towns surrounding him were under the dominion of the Aztec empire, while others were bitterly hostile to it.

Of course, his arrival did not go unnoticed by the Aztecs themselves. Although the capital of the empire, Tenochtitlan, was far inland, the Aztec king Montezuma was well aware of the events in his kingdom. Montezuma was however uncertain about how he should react to the Spaniards—Cortés had arrived in a year of great prophetic uncertainty and dread on the Aztec calendar, a year when the feathered serpent-god Quezalcoatl could potentially return. Now Montezuma received reports of strange men from the East—their appearance and dress were very unusual, and they had metal tools that roared and destroyed things at a distance. Some of them could even merge with strong, exotic beasts (the Aztecs had never seen horses before, and a Spaniard seated on a war-horse was initially mistaken as one creature).

Had Montezuma mobilized his forces against the Spanish at this point, it's doubtful that Cortés could have survived. Although the Spaniards had superior weapons, their body armor was still penetrable by arrows, and Montezuma had tens of thousands of archers and other warriors at his disposal. Yet, he hesitated—and this decision cost him both his empire and his life.

Once Vera Cruz was established and Cortés had an understanding of local geography, he set off for Tenochtitlan. The capital was over 200 miles distant, and it took Cortés 3 months to make the journey. Along the way, they encountered many towns and villages. Some of these were unwilling subjects of the Aztecs, and the Spanish convinced them to rebel. Others were overt enemies of the empire, and agreed to fight alongside the Spanish against their longstanding foe.

Some of the latter were not easily persuaded. The city-state of Tlascala rejected messengers from one of Cortés' native allies, telling them, "Now we are going to kill those whom you call *Teules* [gods] and eat their flesh. Then we shall see whether they are as brave as you proclaim. And we shall eat your flesh too, since you come here with treasons and lies from that traitor Montezuma."[5] The Tlascalans were a fierce people, and they fought a three-day battle with Cortés and his men. Bernal Díaz, one of Cortés' soldiers, wrote that the last day was the most desperate: "We were four hundred, of whom many were sick and wounded, and we stood in the middle of a plain six miles long, and perhaps as broad, swarming with Indian warriors. Moreover we knew that they had come determined to leave none of us alive except those who were to be sacrificed to their idols. When they began to charge, the stones sped like hail from their slings, and their barbed and fire-hardened darts fell like corn on the threshing-floor, each one capable of piercing any armor or penetrating the unprotected vitals. Their swordsmen and spearmen pressed us hard, and closed with us bravely, shouting and yelling as they came."[6]

Nevertheless, the Spanish prevailed. They finally convinced the Tlascalans that they truly were marching on the imperial city to capture it, and so the Tlascalans joined their cause. Cortés thus accumulated many allies as he marched toward the Aztec capital.

Montezuma monitored the advance of the Spaniards, and tried to dissuade them along the way with bribes. As Cortés worked his way inland, Montezuma sent several sets of messengers with gold and other gifts, both to show his friendship and also asking Cortés not to proceed any further. The Aztecs were surprised by the Spaniards' reactions to the gifts. One Aztec later remembered:

"The Spaniards appeared much delighted—they seized upon the

gold like monkeys. Their faces flushed, for clearly their thirst for gold was insatiable. They starved for it, they lusted for it, they wanted to stuff themselves with it as if they were pigs."[7]

For their part, the Spanish were horrified and disgusted with the brutal native culture. Díaz described the human sacrifices which he and the rest of the Spanish often witnessed:

"They strike open the wretched Indian's chest with flint knives and hastily tear out the palpitating heart which, with the blood, they present to their idols in whose name they have performed the sacrifice. Then they cut off the arms, thighs, and head, eating the arms and thighs at their ceremonial banquets. The head they hang up on a beam, and the body of the sacrificed man is not eaten but given to the beasts of prey."[8]

Wherever possible, the Spanish made it a point to halt the human sacrifice, cannibalism, and other abominations in the towns they encountered.

The Spanish entered the Valley of Mexico in November of 1519. They were astonished at the imperial capital of Tenochtitlan—a large island city in the middle of a vast lake. Montezuma actually welcomed the Spaniards personally as they entered, and bid them to be his guests. This was a grave mistake.

The Spanish had already been warned by their allies that the Mexicans planned to kill them once they were in the city. In addition, they immediately realized that the city was a natural fortress, and that they would be unable to capture it in a conventional battle. Unconventional tactics would be necessary. Therefore, after staying in the city for a short while as Montezuma's guests, they took the king prisoner and hostage in his own palace.

Cortés demanded both obedience and "tribute" from the king and all his vassals. So Montezuma called on all the regions under his control to pay tribute to the Spaniards. Gold and other valuables began to arrive at the capital, in addition to the vast amount already stored in the royal treasury. This caused considerable dissension among the Aztecs. Some flatly refused to pay tribute to the Spanish—others among the nobility went further and, the king being incapacitated, began to plot to assume the throne themselves. Nevertheless, the Mexican empire was effectively paralyzed for several months with the Spanish in their capital.

Meanwhile, the Spanish were dividing up the spoils—and arguing over it as well. Many of Cortés' men had noticed that the treasure they had captured was dwindling rapidly, as Cortés and other leaders of the expedition helped themselves to it. Cortés was forced to give a speech that was all "honeyed words," promising the men that everything would be split up fairly—promises that were recognized as empty even as he spoke them. In describing this scene, Díaz captured the flavor of the moment by noting dryly that "all men alike covet gold, and the more we have the more we want..."[9] His suspicions, and those of the other men of the expedition, were correct. Ultimately, the common soldiers would receive very little of the incredible wealth they had captured together.

With the king as prisoner, the Spanish were safe for the moment, but were at a loss what to do next. An unexpected event forced Cortés' hand: he received word that an expedition of 19 Spanish ships had arrived at the coast. This expedition was led by Panfilo Narvaez, and had been sent by Diego Velázquez with 1,400 soldiers to arrest Cortés and bring him back to Cuba to stand trial for his rebellion.

In a breathtakingly daring move, Cortés split his forces. Leaving only a small force in the Mexican capital, he took several hundred men back to the coast. He met Narvaez's men as they marched inland—a substantially larger force than his own—and brilliantly outmaneuvered and defeated Narvaez in a short battle. Cortés then persuaded Narvaez and his men to join him, and they all marched back to Tenochtitlan—where very bad news awaited them.

Cortés had left a lieutenant, Pedro de Alvarado, in charge of the small garrison. Alvarado had apparently committed a massacre among the Mexicans during one of their religious festivals. The enraged Aztecs had besieged Alvarado's men in their quarters, and a full revolt against the Spaniards was in progress when Cortés arrived. The Aztecs allowed Cortés to join his garrison, but it was a trap. They resumed fighting once the Spanish were surrounded again.

After several days of this, Cortés asked Montezuma to ask his people to allow the Spanish to leave the city. Montezuma went on a rooftop and tried, but was killed by several stones from his own people.

The Spaniards now realized that their only chance of survival was to escape from the city. So they tried to slip out over a causeway in the dead of the night. They were detected, however, and thousands of Aztec warriors descended on them. Many people died in the desperate battle—some of the Spanish drowned after falling into the water, weighted down with the gold they had stuffed into their pockets. Nevertheless, Cortés escaped with most of his men.

This was Cortés' darkest hour. His men were defeated, wounded, starving, and many were ill. To make matters worse, most of their weapons, including all of the muskets and artillery, had been lost. But

Cortés didn't falter. He rallied his men, and through the end of 1520, built an alliance of anti-Aztec forces among the various peoples of the region. He eventually led them in an attack against Tenochtitlan, which finally fell after a long battle on August 13, 1521. Thus, the Spanish conquered the mighty Aztec empire.

Many have criticized the Spanish for this act of conquest, portraying it as a naked act of aggression. Others have said that the Spanish only wanted to Christianize the native peoples. Perhaps the best summary comes from one of the conquistadors themselves: he wrote that he and his companions "came to Mexico in the service of God and His Majesty, and to give light to those who sat in darkness—and also to acquire that gold which most men covet."[10]

Cortés' capture of the mighty—and fabulously wealthy—Mexican empire resulted in his vindication back in Spain. King Charles V named Hernán Cortés as governor of "New Spain," i.e. the Spanish possessions in the Americas. Cortés' success was widely proclaimed, and the wealth of the conquered territories inspired other men to try for the same in the yet-unconquered regions. One of those men was Francisco Pizarro, who set his eyes towards the region that we now call Peru…

* * * * * * *

After the victory in Peru, when a priest asked him to do more to convert the natives, Pizarro's response was, "I have not come for any such reasons. I have come to take away from them their gold." [11]

Hernán Cortés had set out into unexplored territory, lured on by rumors of a vast empire with untold wealth. He brought 400 men (later supplemented with another thousand or so), and fought an empire made up of millions of people. As noted earlier, it seemed an act of breathtaking bravery—or blind stupidity. Nevertheless, he won anyway.

Pizarro also set out into unexplored territory. He also learned of, and then set out to fight, an empire made up of millions of people.

He brought 168 men.

Pizarro had learned well from Cortés' experiences. When he arrived in what today is Peru, he found a vast empire—the Incas—enmeshed in a bloody civil war. The Inca king Huayna had died around 1526, and had left no clear line of succession. Two of his sons, Atahualpa and Huascar, had each claimed the throne, and were battling over it.

Pizarro brazenly marched directly to the town of Cajamarca, next to the encampment of Atahualpa's army. Atahualpa had just captured his brother Huascar in battle, and was curious about the strangers. He perceived no threat from the tiny group of Spaniards, and agreed to meet with them the following day.

On the afternoon of November 16, 1532, Atahualpa entered the town square of Cajamarca, "carried aloft on his palanquin by eighty nobles and accompanied by a redoubtable host of several thousands of his people. Seated in majesty at the center of the huge square, he contemplated the small handful of men who had managed to penetrate his domains."[12] But Pizarro gave a signal and suddenly a cannon, hidden earlier in the day, was exposed and fired directly into the crowd. Other soldiers and horsemen also leaped out of concealment and began to massacre the defenseless and now panic-stricken mob. At the same time, Pizarro and a few of his men leaped directly onto Atahualpa, taking him prisoner. The crowd of terrified people, still being fired on by the Spanish, trampled each other in their efforts to escape. More than two thousand Incas were killed in the massacre, with no losses to the Spanish.

Pizarro demanded a tremendous ransom from Atahualpa. He touched a spot high on the wall of the room in which Atahualpa was being kept prisoner (as high as he could reach, almost nine feet), and required the king to fill the *entire room* with gold and treasure to that point. The king ordered his people to comply. As one author writes:

> "The amassing of the Inca's treasure was one of the most emblematic acts in the history of all empires. It displayed to perfection the obsession of Europeans with the wealth associated with precious metals. Above all, it displayed their complete indifference to the destruction of the cultures with which they came into contact. As the ornaments were rounded up by the Inca's messengers from the four corners of his part of the empire—plates, cups, jewellery [sic], tiles from temples, artefacts [sic]—they were systematically melted down under Spanish supervision, and reduced to ingots. Over those four months from March to June 1533, bit by bit the artistic heritage not simply of the Incas but of a great part of Andean civilization disappeared into the flames. For two thousand years the craftsmen of the Andes had applied their techniques to working and decorating with gold. This became no more than a memory. At Cajamarca alone the Spaniards managed to reduce the ornaments to 13,420 pounds of gold and 26,000 pounds

> of silver. In subsequent weeks, they came across equally fabulous treasures, which were likewise consigned to the furnaces."[13]

As it turned out, the Inca king should have rejected the Spanish demand for ransom. In a shameful and criminal act, Pizarro had him executed anyway.

Pizarro had effectively beheaded the Inca empire, and controlled much of it, but had not yet fully conquered it. That would not happen for another 35 years, after a series of battles with various groups of Incas and even a civil war among the Spanish themselves. Nevertheless, the Spaniards wasted no time in plundering their new territories. Guaman Poma, a native who was raised under Spanish rule, wrote caustically about Pizarro and the other Spaniards:

"They did not wish to linger a single day in the ports. Every day they did nothing else but think about gold and silver and the riches of the Indies of Peru. They were like a man in desperation, crazy, mad, out of their minds with greed for gold and silver."[14]

One of Poma's drawings depicts an Inca named Huayna Capac seated across from a conquistador. Both are pointing to the same golden vessel. The Inca asks in his own language (Quechua), "Cay curitacho micunqui?" (Do you eat this gold?) The Spaniard, uncomprehending, replies in Spanish, "Este oro comemos." (We eat this gold.)[15]

Eventually, the Spaniards would erect an efficient system all over Central and South America to plunder their new possessions. New mines were dug, and old ones were expanded. Untold numbers of slaves, both native and imported from Africa, labored and died in the mines of the Americas. And so a vast river of gold and silver bullion began to flow each year across the Atlantic Ocean, to Spain.

The Aztecs and the Incas were proud civilizations. They built impressive cities (some at high altitude), had sophisticated farming and irrigation systems, and gathered extensive knowledge of the calendar and astronomy. They did all of this despite never having invented the wheel or the arch, and having no large draft animals. The Incas especially were noted for ruling and administering a vast empire (stretching from modern-day Ecuador to Chile and Argentina), without even having a written language! Yet, in the wars over the gold of the Americas, these nations had lost.

And now other nations would join the fray...

* * * * * * *

For six hours the English and Spanish ships maneuvered in the narrow confines of the harbor of San Juan de Ulua, pounding each other with their cannons at point-blank range.

"Amidst the cannonade [Sir John] Hawkins courageously cheered up his soldiers and gunners and called to Samuel, his page, for a cup of beer, who brought it to him in a silver cup; and he drinking to all men, willed the gunners to stand by their ordnance lustily like men. As soon as he had finished his beer and set down the cup it was carried away by a shot, at which he exclaimed, 'Fear nothing, for God who hath preserved me from this shot will also deliver us from these traitorous villains.'" [16]

The wealth sent from the Americas to Spain was staggering. One author notes, "Gold was the first great lure offered: Columbus, Cortés, Pizarro, and every subsequent adventurer placed the search for gold at the head of his priorities. The Caribbean, where Columbus had seen natives eat off plates of gold, was the primary producer; the precious metal was in the early days panned from mountain streams. In the first two decades of the sixteenth century the Spaniards probably collected around fourteen tons of gold (14,118 kilos) from the Caribbean. The news of the discovery of gold in Peru led to further exploration, discovery, and exploitation. Most of the metal went to Spain, where it excited astonishment. An official of the emperor's treasury wrote from Seville in 1534 that 'the quantity of gold that arrives every day from the Indies and especially from Peru, is quite incredible; I think that if this torrent of gold lasts even ten years, this city will become the richest in the world.'"[17]

Of course, this "torrent of gold" did not go unnoticed by the other nations of Europe. They saw no reason why they shouldn't be participating in it as well.

One small event in 1545 foreshadowed much larger trends in the future. England was at war with France, and English sea-captain Robert Reneger had captured a French vessel and brought her into a neutral (Spanish) port. The Spanish claimed that some of the goods aboard the French ship were actually Spanish property, and seized the entire ship. Reneger vowed revenge, and later that year, ambushed the Spanish treasure ship *San Salvador* off the coast of Hispaniola in the Caribbean. He took a tremendous quantity of gold, 124 chests of sugar, and 140 hides.

Reneger claimed he was only taking fair restitution for what had been unlawfully taken from him. The Spanish took a different view, of course. Their anger was not assuaged when Reneger was appointed to a command in the English navy as an apparent reward. This incident set off a series of reprisals (embargoes, confiscations, etc.) between the Spanish and the English, as well as a wave of piracy against Spanish vessels in the English Channel.[18]

Meanwhile, on a more peaceful note, many English seamen were interested in trading with the colonials in the new Spanish territories across the Atlantic. However, Spain had forbidden all such trade. Officially, the colonists in "New Spain" were only allowed to trade with Spanish merchants. The colonials themselves didn't necessarily agree with this policy, however.

So in October 1562 Sir John Hawkins left England on what turned out to be the first of several trading voyages to the Caribbean. He found the colonials eager to trade, as he offered better prices than were available to them under the Spanish monopoly. His trip was successful, and very profitable.

In October of 1564, Hawkins made another voyage, this time with the official backing of Queen Elizabeth. He found that the colonials had been chastised for breaking the trade monopoly two years earlier—but they still wanted his goods. Thus, at Borburata the Spaniards wouldn't grant him a license to trade until Hawkins had "threatened" the use of force—so they could tell their King they had had no choice in the matter. The rest of Hawkins' trip unfolded in similar fashion:

"After obtaining hides and beef at Curacao, Hawkins moved on to Rio de la Hacha where there was a further comedy. The treasurer, Miguel de Castellanos, hesitated about nodding his approval until the English should threaten to set fire to their houses 'in order that they might take various depositions of witnesses and prove that they were forced to trade with him.'" Once the "threat" was out of the way, enthusiastic trading then took place. "They were so delighted at Hawkins' initiative that they booked firm orders for his next voyage and the treasurer even issued him with a testimonial praising the behavior of his men."[19]

When King Philip II of Spain heard of this farce, he was not impressed, and made his displeasure known. Thus, when Hawkins returned to the Caribbean in 1568, he now found the colonists, or at least their officials, unwilling to grant him a license to trade. The governor of Borburata officially rebuffed Hawkins, but then looked the other way while he and his men traded with the colony. However, Hawkins received a much more hostile reception elsewhere.

At Rio de la Hacha, treasurer Miguel de Castellanos had been reprimanded for his earlier friendliness to the English, and this time he refused to allow the English to even refill their water casks. This resulted in one of Hawkins' ships, captained by Francis Drake, putting a cannon shot through the treasurer's house on shore. Hawkins eventually landed soldiers and captured the town, burning a few buildings in the process. (This was the first attack on a Spanish colony in the New World by English privateers.) The colonists themselves then demanded that the treasurer allow the English to trade, which permission was reluctantly granted.

After Rio de la Hacha, Hawkins visited a few more ports, but soon turned back towards England. However, while still in the Caribbean his fleet was hit by a heavy storm, and he needed to make repairs before risking the Atlantic crossing. The only suitable location was San Juan de Ulua, the port of Vera Cruz on the mainland...and the port used by the heavily-armed Spanish *flota* (treasure fleet).

Hawkins had little choice, so he decided to risk it—get into port, make his repairs, and get out before the *flota* arrived. He made it into port by pretending to be Spanish (which was made easier by the two Spanish ships he had found and forced to accompany his fleet). By the time the Spanish realized what was happening, he was in the harbor, and had captured an island embankment of shore guns as well.

Hawkins assured the governor, Antonio Delgadillo, that his intentions were entirely peaceful—to repair and re-provision, and be on his way. The governor agreed to this arrangement. Unfortunately for Hawkins, the Spanish *flota* showed up the next day—eleven large ships, and two men-of-war.

"To add further tension to an awkward situation, aboard the [Spanish] admiral's ship was the viceroy of New Spain, Don Martin Enriquez, coming out to take up his appointment, under strict instructions from King Philip to defend Spain's colonial trade from English interlopers... Hawkins was placed in a fearful dilemma: if he allowed the Spaniards to enter San Juan de Ulua he would be completely at their mercy, for he was outclassed in numbers and in armament, and while the Spanish fleet was trim and in first-class condition, the English vessels were battered, [his flagship] a lame duck. Yet if he refused them entry so that they had to anchor outside their own harbour, where they would be at the mercy of the winds, he would be perpetrating a diplomatic incident that might well lead to open war in Europe, for he was still 'the Queen's Officer'."

Don Martin thought about forcing an entry, but Hawkins held the shore batteries. The two men finally came to an agreement: the Spanish would come into the harbor, but Hawkins and his ships would be unmolested. Little did Hawkins know that the viceroy had no intentions of keeping the truce…

The Spanish fleet came quietly into the harbor, and all seemed well. But all that night, Don Martin secretly made preparations. Hawkins became suspicious and sent a messenger, Robert Barrett, to protest—who was immediately clapped into irons by the Spanish. At dawn the shore batteries were attacked by Spanish soldiers, while the Spanish ships (who had maneuvered into position) opened fire at the English vessels at point-blank range. Hawkins was outraged at this act of "treachery, unbecoming of a gentleman." He and his men fought bravely, sinking several ships (including the viceroy's flagship, although Don Martin himself survived).

Nevertheless, the 6-hour battle left Hawkins fleeing the harbor with only one other ship besides his own. The other ship, commanded by Francis Drake, made it home with only 65 survivors. Hawkins' ship had a mere 14.

Peaceful trade with the Spanish was now shown to be impossible. English anger was stoked further when Hawkins' messenger Robert Barrett, who had refused to recant his Protestantism before the Spanish, was brought back to Spain and burned alive in the public square of Seville. Francis Drake vowed revenge for the treachery of Don Martin Enriquez—from now on, he would plunder the Spanish treasure fleets and "singe the King of Spain's beard."

Drake and a growing number of other sea-captains were now sent out to prey upon the Spanish fleets. Although Drake's motive was, at least in part, revenge, his Queen had other goals in mind. As one author notes, "Gold was the grand object of trade itself, and the close interest of the crown in every major commercial project was clearly inspired by hopes of gaining or saving bullion."[20]

Queen Elizabeth began sponsoring a series of voyages, to relieve the Spanish settlements and ships of their gold. Commissioned by the Crown, these men were "privateers" rather than pirates. (The Spanish, of course, saw no such distinction.) Drake in particular was spectacularly successful in capturing gold for his Queen—his voyage of 1577-1579 alone brought back more than ten tons of gold, silver, and jewels from the Spanish (causing the Queen to knight him upon his return).

Drake and the other privateers did more than just steal from the Spanish—Drake was the first Englishman to circumnavigate the globe, followed a few years later by Thomas Cavendish. Nevertheless, their plundering put more and more strain on the already tense relationship between England and Spain, until open war finally broke out in 1586.

The war unleashed a horde of marauders against Spanish shipping. In peacetime, there had been many small pirates raiding merchant ships (of every nationality) in the English Channel—these were now offered amnesty and privateering licenses from the Crown against the Spanish.

In addition, the activity of the Queen's current privateers was stepped up. The *Santa Ana*, whose story began this chapter, was attacked by Sir Thomas Cavendish off the coast of Baja California on November 14, 1587. This was one of the "Manila galleons," so named because they transported gold and other treasures to the Spanish Philippines, and brought back silks, spices, and other wonders of the Orient. Captain Tomás de Alzola fought gallantly for six hours, repelling several English attacks, but finally was forced to surrender his sinking vessel to Cavendish and his men. The *Santa Ana* carried 600 tons of rich merchandise, including hundreds of thousands of gold pesos—ten times as much as the two English ships could carry. Cavendish put the prisoners ashore, loaded his ships with as much gold and treasure as they could hold, and burned the rest.

The privateers did not limit themselves to attacking ships at sea. In the first few days of 1586, Sir Francis Drake captured and ransacked the Caribbean city of Santo Domingo. He burned more than a third of it before the Spanish agreed to pay his demands for ransom (25,000 ducats in gold). A month later, he did the same to Cartagena—"perhaps the most important city in the Caribbean"[21]—this time for 107,000 ducats. Numerous other towns and forts were attacked as well: Vigo Bay, Santiago, San Agustin.

Drake's raids were not only in the Americas. On April 29, 1587, he attacked the port of Cadiz—on the coast of Spain itself. He captured or destroyed 24 ships in less than one day.

By this time, the Spanish had had enough of these provocations. Decades of English depredations had left King Philip II determined to crush England once and for all. Thus, on May 28, 1588—after several years of preparation—an enormous fleet of Spanish ships set sail against England. 130 ships, carrying a total of 18,973 troops and 2,431 cannon—the mighty Spanish Armada.

* * * * * * *

"All wars and affairs afoot today are reduced to this one enterprise."
Don Juan de Idiáquez

The Spanish Armada was actually just one half of Philip's overall plan. The other half was the Army of Flanders—the most feared army in Europe.

This Army was in the Spanish Netherlands, under the command of the Duke of Parma. Parma's men had been fighting against the Dutch rebellion since 1581. They were veteran, crack troops—considered by many to be the best soldiers in Europe. 17,000 of these soldiers were selected to invade England under the Duke's command.

King Philip's plan had several steps. First, the Armada was to sail up into the English Channel, fighting off whatever ships England sent against it. Once it was in position off the Flemish coast, the Duke of Parma was to embark his soldiers onto boats and cross the Channel under the protection of the Armada. His men were to land in Kent, along with artillery, supplies, and other troops from Armada transports—and then march directly to London, to capture it as quickly as possible.

Some have criticized Philip's plan as foolhardy—surely a landing force totaling 23,000 men would be insufficient to subdue an entire country. In reality, his plan was quite feasible. Outside of her navy, England had very few men under arms at that time—a census in July 1588 of all "martial men" (i.e., veterans) in England produced a list of only 100 names.[22] Local militias had little in the way of weaponry other than swords and bows—little match for Spanish muskets and cannon. Many English towns were fortified, but few of the fortifications were strong enough to withstand modern artillery. In addition, the distance from northeastern Kent to London was only 80 miles—and in 1592 Parma marched his men 65 miles in 6 days through Normandy, fighting against numerically superior forces along the way.[23] These were the same men slated to invade England—and a fearsome force they would have been.

Unfortunately for Philip, the battle did not go according to his plan. The English navy clashed with the Armada in the Channel, sinking several ships. More importantly, the Armada was never able to rendezvous with Parma (the messengers sent to him were delayed in arriving. By the time his troops were ready to load up into their barges and boats, the Armada had been pushed too far north by the winds and the English.)

The Spanish naval commanders debated about whether they should push back into the Channel and try again to meet up with Parma. Their fleet was battered, but still strong, although supplies and ammunition were low. Unanimously, they decided to return—but three days of stiff winds from the southwest merely pushed them further north. Finally, they acknowledged defeat, and turned for home.

Going back through the Channel was impossible, so they turned west, to pass the northern coasts of Scotland and Ireland, circle around into the Atlantic, and then back to Spain. This was not an uncommon route for ships to take. Unfortunately for the Spanish, as they were swinging around Ireland two Atlantic depressions also arrived, merging into one tremendous storm on September 21st—and the Spanish ships were right in its teeth. An English official in Ireland described it as "a most extreme wind and cruel storm, the like whereof hath not been seen or heard a long time, which put us in very good hope that many of the ships should be beaten up and cast away upon the rocks."[24] That's exactly what happened: the ships of the Armada were scattered, thrown about, and smashed onto the rocky coastlines of Ireland and Scotland.

The survivors of the storm struggled home as best they could, arriving one by one. Food rations had been cut as soon as the decision had been made to return via the northern route, but that wasn't enough—over the 45 days it took to return to Spain, many of the survivors had only received 30 days' worth of rations. In addition, fresh water ran out for many ships before they made it back. Almost every ship had four or five men dying every day.[25] Some ships managed to return but then wrecked along the Spanish coast, collided with other vessels, or sank, because the crews were too weak to man the sails or pump out seawater.

Of the 130 ships that left Spain, only 60 returned. 15,000 men were lost. As the magnitude of the disaster unfolded in Spain, King Philip was horrified. He moaned, "Very soon we shall find ourselves in such a state that we shall wish that we had never been born." The Duke of Medina, the commander of the Armada, wrote to the king: "I am unable to describe to Your Majesty the misfortunes and miseries that have befallen us, because they are the worst that have been known on any voyage; and some of the ships that put into this port have spent the last fourteen days without a single drop of water." The Duke's majestic flagship had been so severely damaged that "she had to be cinched with three great hawsers wrapped round her to prevent the seams from opening. And thus, literally held together with string, she returned."[26]

The people of Spain were stunned by the catastrophe. As one author notes, "Even survivors could not always shed much light on what had happened: one man from the *Trinidad Valencera* was still so traumatized by his experiences that 'it was difficult to understand what he was trying to say'."[27] One Spaniard wrote that it was "the greatest disaster to strike Spain in over six hundred years." Another wrote that it was "worthy to be wept over for ever…Almost the entire country went into mourning. People talked about nothing else."[28]

The Armada was made possible by American gold and silver—Spain spent enormous amounts of money on this and other military operations, and could not have afforded to do so otherwise. Even the Duke of Medina, commander of the Armada, had recognized this: when asked if he wanted additional galleons for the Armada, to be taken from their escort duties for the treasure fleet, he had refused. He said, "The link between the two continents is the foundation of the wealth and power we have here."[29]

However, American gold and silver also helped Spain to recover fairly quickly from the Armada disaster. Philip II even tried again, sending fleets against England in 1596 and 1597—but they were driven back by storms.

Meanwhile, the privateers were on a rampage. Just from 1589-1591, English privateers captured some 300 Spanish merchantmen, worth about £400,000. From 1585 to 1603, even excluding Francis Drake, there were 74 separate English expeditions, composed of 183 individual voyages. This free-for-all continued until 1604, when peace between England and Spain was finally achieved.

No doubt the Spanish were relieved that the war with England was over, thinking that their problems with piracy were over. In fact, in many ways their problems had just begun…

* * * * * * *

> *"Once the English had [captured and] boarded her, gone below, and inspected the chests, coffers and bales, it was as if they had been transported to Aladdin's cave. There were diamonds, chains encrusted with jewels and gold objects of rare workmanship; they feasted their eyes on crystal-ware garnished with gold and pearls, gold cutlery set with precious stones, elaborate collars, 'strings of pearls orient', gold buttons, rings, and bracelets. Then there were considerable quantities of pepper, cloves, nutmeg, ginger and less common spices, drugs like camphor, benjamin and frankincense, musk and other perfumes, Chinese silks, damasks and taffetas, curled Cyprus cloth, Indian calicos and lawns of many varieties, carpets, quilts and hangings of rare design and luxuriant texture…elephants' teeth [ivory], porcelain, vessels of China, coconuts, hides, ebony wood as black as jet, bedsteads of the same, cloth of the rind of trees, very strange for the matter and artificial in workmanship…"*
>
> *On the capture of the treasure galleon* Madre de Dios*, taken in the Azores in 1592.*[30]

Sir Francis Drake and others had shown that privateering could be very, very profitable—and many were attracted to it as a result. But the *Madre de Dios* truly captured the public's imagination. Here, in one ship, was a staggering amount of gold and other treasures.

It's difficult for us to comprehend the amount of wealth that Spain was gathering from the Americas. "According to one authority, the total European stock of gold and silver at the end of the 1500s was nearly *five times* its size in 1492. The volume was so enormous that the armed convoys that transported the treasure to Europe averaged about sixty ships; on occasion, the convoys included as many as one hundred ships. Each of these vessels carried over two hundred tons of cargo in the 1500s and around four hundred tons on larger ships in the 1600s. In 1564 alone, 154 ships arrived at Seville to debark their cargo of treasure."[31] It's easy to see the tremendous temptation that other nations faced—and the incentive for young men to take to the seas and get some of that treasure for themselves.

Even before the war with Spain, England had not been the only nation engaging in privateering—other nations participated as well, with the French Huguenots being especially feared by the Spanish. Thus, although "official" English privateering was mostly halted once the war ended, that of other nations continued—especially by the Dutch.

The Dutch had been fighting against the Spanish in the Netherlands for decades (part of what is now known as the Eighty Years' War: 1568-1648). Indeed, the Spanish war against England had been fought in part because of English support of the Dutch. In 1609, Spain finally agreed to a truce in the Netherlands—and the Dutch were free to take to the seas in earnest, establishing two great trading corporations to facilitate worldwide trade—and privateering.

Dutch privateering introduced several new innovations. Previously, the Pacific coast of South America was mostly ignored by privateers, but the Dutch spent much of their time raiding along it. It was only lightly defended, and the Dutch made the most of this situation (although they certainly didn't ignore the riches of the Caribbean either).

Also, privateers to this point had only been concerned with raiding—if they captured a town at all, it was only to hold it for ransom. The Dutch, however, started capturing territory and keeping it—such as the island of Curacao. They even captured a large portion of Portuguese Brazil for a number of years (until it was taken back).

The Dutch turned out to be excellent privateers—in 1627 alone the Dutch West India Company claimed to have captured 55 Spanish and Portuguese ships. A Dutch admiral even pulled off what is perhaps the single greatest act of privateering in history—in 1628, Piet Heyn captured the *entire* Spanish treasure fleet for the year: fifteen ships full of gold, silver (177,537 pounds of it!), silks, and other treasures.

Of course, the Dutch were not the only problem the Spanish faced. Hostile relations with England waxed and waned, and in 1655 the English captured the island of Jamaica. The English soon turned it into a haven for pirates of all sorts: a "commission" from the governor was available for just about any venture against the Spanish, in exchange for a cut of the booty once the pirates returned to port.

Also by this time, the "buccaneers" of the Caribbean were becoming established. Originally the term referred to a motley group of marooned sailors, escaped indentured servants, and other misfits on the northern part of Hispaniola. They lived largely off the feral cattle in the region, cooking the meat on a metal grate called a *bouca*—thus, the French called them *boucaniers*. Over time, however, they also took over the island of Tortuga, and began actively trading with the colonies around them. Once they were seaborne, piracy wasn't too difficult to contemplate—and afterwards they could put in at Tortuga, Jamaica, or other pirate havens, to sell their stolen goods (and buy some strong drink).

The Spanish were of course unhappy about these pirate bases in their midst. Rumors began to circulate that the Spanish were about to invade Jamaica, to shut down its pirates. The rumors were actually false, but the governor of Jamaica believed them, and decided to do something about it. He commissioned a man who has become one of the most notorious among the long, bloody list of the buccaneers: Henry Morgan.

Morgan, in a "pre-emptive" strike against the Spanish, landed in Cuba and sacked the city of Puerto Principe. Next, he went to Portobelo in northern Panama, and took it as well. He used captured nuns and friars as human shields during his attack, and sacked the city for some time once it was in his hands, torturing his prisoners. After returning to Jamaica with his booty, and now rampaging out of control, he attacked other Spanish settlements as well, culminating in his infamous assault against the city of Panama with 1,800 men in early 1671. The city fell quickly to his attack, and he spent the next four weeks torturing, maiming, and mutilating the inhabitants in an effort to force them to tell him where all their treasures had been hidden.

The Spanish were predictably outraged by Morgan's attacks, especially as the Treaty of Madrid had just been signed with England a few months earlier—a treaty in which Spain accepted the English claim to Jamaica and other territories, in return for an end to hostilities. Morgan was therefore arrested and brought back to England, but nobody was surprised when he received no punishment. Indeed, he was appointed as deputy governor of Jamaica three years later, in 1674.

Morgan was in many ways symbolic of the new wave of piracy. Privateers like Sir Francis Drake had operated with the explicit (or tacit) approval of their governments—but Morgan had attacked Panama despite explicit orders not to do so. Drake and others of his time had for the most part treated their prisoners honorably—Morgan and the other buccaneers were often savage, even unnecessarily sadistic, to theirs (assuming they left any alive at all). Drake and others had attacked the shipping of other nations (usually Spain) to which their governments were hostile in various degrees—but the new buccaneers for the most part were not loyal to any particular country. There were many "freebooters" who acknowledged no governmental authority of any kind, freely attacking any hapless vessel or settlement that they came across. Some buccaneers attacked shipping of only one nationality (which wasn't always Spanish—some pirates exclusively attacked English shipping), but this was usually out of spite or a grudge rather than a political choice.

This period gives us many of the famous pirates of history. The notorious Edward Teach, known as Blackbeard—a cruel man known for wearing burning fuses in his dreadlockish hair, and whose crew thought to be possessed by a demon. Bartholomew Roberts, known as Black Bart—who preyed mostly on English and French shipping, and was wildly successful at it (capturing *400 ships* by 1721). William Kidd, known today as "Captain Kidd"—originally sent out as a pirate hunter, then turned pirate himself, only to be arrested and hanged for his crimes once he returned to port.

Increasingly, the buccaneers found their welcome in various ports waxing and waning with political trends in Europe. When England was fighting the Spanish, or the French, or the Dutch, then the pirates would be more or less welcomed into English ports, as long as they had been out attacking enemy shipping. (The same was true for French and Dutch, and to a lesser degree Spanish, ports.) But over time, as the buccaneers became more and more notorious for the viciousness of their crimes, the European nations eventually agreed among themselves that this scourge had to be stopped. Over a period of years, the pirates were eventually all caught and exterminated. Singular acts continued here and there, but the era of the buccaneers was over by 1730 or so.

Still, year after year, the bullion continued to flow across the Atlantic to Spain. (By this time, it was mostly silver from the mines of Central and South America, rather than gold.) Yet, starting around the turn of the eighteenth century, Spain's ships were now unmolested for the most part by the other European nations (the occasional capture of a ship during wartime notwithstanding). Why did these nations allow this?

Simply because they now recognized it was in their best interest to do so. Most Europeans realized that the bullion that flowed into Spain immediately flowed out again in trade to the other countries of Europe. In addition, Spain's colonies in the New World were by this time trading openly and freely (although illegally) with other nations.

The Europeans realized that the status quo was much more profitable than war. For example, in January of 1739, the prime minister of England explained to his Parliament that, "It is true all that treasure is brought home in Spanish names, but Spain herself is no more than the canal through which all these treasures are conveyed all over the rest of Europe."[32]

Over time, Spain eventually lost her colonial possessions, and the flow of bullion stopped. Although it lasted several centuries, Spain did not emerge from this experience as a wealthy nation—indeed, she had reneged on her debts several times during this period. Part of this can be explained by the ruinous series of wars that she always seemed to be involved in: the Thirty Years' War in Germany, the Eighty Years' War with the Dutch, a series of wars with England, and so on. No doubt she could have done very little of this fighting were it not for the wealth of the Americas. Thus, in a sense, these wars in Europe were also part of the larger war for gold.

And so, the greatest war over gold in history came to an end. Nations had fought over it, entire peoples were enslaved and labored to produce it, and armies and armadas fought massive battles because of it. Indeed, the effects of these battles linger on today, as treasure hunters search eagerly for the numerous sunken Spanish treasure galleons which have remained undiscovered. Hundreds of years after the war, the gold lies quietly today on the ocean bottom—as pristine and gleaming as the day it went down, and waiting patiently to be found and admired once again.

CHAPTER 3

Nazi Lust for Gold

On April 4, 1945, nearing the end of World War II, two German officers were overtaken by American forces on the road between Merkers and Erfurt, Germany. Fearing for their lives, they fled into the woods. One of these men, Albert Thoms, was an official at the German central bank (the Reichsbank*). Within a week Thoms would be captured by the U.S. Army and assisting the Allies.*

The Allies discovered the Nazis had stolen nearly $600 million worth of gold from the central banks of France, Belgium, Holland and other countries, as well as over $50 million worth of gold from the homes, hands and mouths of the millions of Jews, gypsies and homosexuals that were put to death in the Nazi concentration camps.

The Nazi plunder of Europe shows that, far from being a "barbarous relic," gold is as eagerly desired today as it ever was.

At the beginning of World War II, in 1939, the majority of European countries still kept large reserves of gold bars and bullion in their central banks. At the time it was believed that Germany had approximately $100 million worth of gold reserves in the *Reichsbank.*

It soon became apparent that Germany's gold reserves had been spent—on weaponry, supplies and so forth; and yet the Germans, now controlling large parts of Europe beyond their borders, continued to make payments to its suppliers in Spain, Portugal, Turkey and Italy.

It also soon became clear that the German forces moving through Europe were swiftly followed by a crew of financial vultures, taking the gold reserves and artworks of each country that had been overrun.

The Nazi War Machine

In 1933 Adolph Hitler, the leader of the National Socialist ("Nazi") party, had taken effective control of the government of Germany. At the time he took power, Germany was both economically weak and psychologically impoverished from the loss of land, prestige and face—its fate under the Treaty of Versailles, agreed to at the end of the First World War.

Hitler's fiery speeches about the pride of German people and the German sense of community appealed to the battered ego of the *Deutsche Volk* and his support throughout the country was nearly unanimous.

Hitler recognized the importance of the appearance of legality and formality. Sometimes this meant that he would first have to write a law before he could enforce it, but that was a small inconvenience. After a mysterious fire broke out in the government buildings (the Reichstag) he declared martial law on February 28, 1933, claiming that he feared a Bolshevik uprising. He then set about mobilizing an army with which he hoped to regain that which he believed belonged to Germany.

He sent troops to the Rhineland—an area of Germany that had seen heavy fighting during the First World War. This act was forbidden by the Treaty of Versailles, but the rest of the world was silent on the action.

Hitler deduced that the Allies were easily intimidated. His subsequent invasions of Austria and Czechoslovakia were also militarily unopposed. Finally, the Allies had had enough; when Hitler invaded Poland on September 1, 1939, Britain and France declared war on Germany two days later.

After a quiet winter, in May of 1940 Hitler attacked Belgium, Holland and France. While Belgium and Holland could hardly have been expected to put up much of a fight against such a massive force, France, with support from the British Army, was expected to stop the *Wehrmacht* in its tracks.

The British and French forces actually outnumbered the Germans significantly, although the German equipment was generally better. Unfortunately, they deployed in strength along the French border with Belgium, to meet Hitler's (feinted) attack into that country. This allowed Hitler to bypass them completely by making a surprise assault through the (supposedly impassible) Ardennes Forest, penetrating deep into France. In only six weeks, territory that had been fought over for four years during the First World War fell to the Germans, and the French signed an armistice agreement with the Germans to stop further slaughter. On June 23, 1940, Hitler marched triumphantly through Paris.

The phenomenal victory in France seemed to whet Hitler's appetite for further victory. Convinced that his army could not be defeated he then took the step that would, ultimately, be his downfall: he invaded Russia, reneging on a deal he had reached with Stalin only two years previously.

At first, the attack on Russia, named Operation Barbarossa, appeared to be quite similar to the French campaign. The main problem was getting supplies to his three million soldiers at the front line, as they were advancing so rapidly. Diverting his forces south, and away from Moscow, Hitler missed what was probably his only chance to strike a blow to the center of the Russian mammoth. By the time his troops had finished with Kiev and turned their attention back towards Moscow, it was already getting cold. The Germans were unprepared for the brutal Russian winters, and suffered enormous casualties from the weather. A Russian counteroffensive in December threw the Germans back hundreds of miles, ending with both sides greatly weakened. But Hitler was adamant, and after the spring thaw the Germans resumed their advance.

But 1942 and 1943 went badly for Hitler. He had already lost the "Battle of Britain," his bombing campaign over England, in 1940. In 1942, his overextended invasion of Russia was crippled by the disastrous battle of Stalingrad, beginning in December. He lost over 200,000 men in a failed four-month battle for the city, and never regained the momentum afterwards. From that point on, the Germans were on the defensive in Russia. Then in 1943, the United States entered the war in Europe in the invasion of Hitler's ally Italy. Also in that year, the Allies finally figured out how to effectively fight the dreaded "wolf packs" of German submarines. Instead of sinking large numbers of Allied ships, the U-Boats started to go to the bottom themselves.

Operation Overlord, which resulted in the so-called D-Day landing on June 6, 1944, put Allied troops into France. From there, they swept through the occupied countries of Europe with a speed and ability that had only been matched by Hitler's invading force four years earlier. Over the course of the next year, the Allies pushed the German army back into Germany from both the west and from the east. By April 30, 1945, Hitler, along with his mistress, Eva Braun (whom he had married just the day before), committed suicide in his bunker in Berlin.

As the war drew to a close, the extent of the damage that Hitler had wreaked upon Europe began to be revealed.

German Looting of Central Banks

From the time Hitler ordered his *Wehrmacht* to march into Poland, there were financial moves being made simultaneously to ensure that his war cost the *Deutsche Volk* as little as possible. But the rest of Europe paid dearly.

Most of us are familiar with the tales of brutality and murder inflicted upon the victims of Nazi hatred throughout the Second World War. We are familiar also with the astounding number of victims of every kind during that war: over 50 million lives were lost during those 6 years. The numbers of injured, displaced and psychologically destroyed simply cannot be calculated.

But many are not aware of the degree to which Hitler's forces managed to inflict permanent and lasting financial damage on the countries that he overran. One of his most successful methods of ensuring that the countries that he invaded paid for this dubious honor was to physically loot their central banks and subsequently transfer their gold reserves into the *Reichsbank*, the German central bank, based in Berlin.

According to recent calculations, a total of $580 million worth of central bank gold was taken from these countries. Belgium, one of the smallest (and earliest) countries to fall before the *Wehrmacht*, yielded the largest treasure to the *Reichsbank* coffers—a phenomenal $223 million (by itself more than twice what Germany had when it began the war). The Netherlands yielded another $168 million to the Nazi cause, Italy $64 million, and France $54 million—all in gold, all worth at least 10 times as much in today's values. The balance of monetary gold used to fill the vaults within the *Reichsbank* was obtained from the central banks of Austria, Hungary, Yugoslavia, Czechoslovakia, Poland and Luxembourg.

German Looting of Civilians

On the 28th of November, 1941, a telegram was received by the Secretary of State of the United States. It read:

> By the eleventh decree under the Reich Citizenship law dated November 25, 1941, and effective the following day, all German Jews who have their "usual place of abode" outside Germany or who at any future time transfer their usual place of abode from Germany abroad, are deprived of their German citizenship. All property of such persons as well as of stateless Jews abroad who were formerly German citizens is forfeited to the Reich. The decree states that such property "is to be used for all purposes connected with the solution of the Jewish question."
>
> ...For several years the German authorities have deprived a considerable proportion of the Jewish emigrants of their German citizenships and confiscated such properties as they were forced to leave behind in Germany but the present decree makes this practice universal and automatic. Since it apparently does not apply in the so-called Government General or in the German-occupied territories of the Soviet Union, these areas will apparently be considered as "outside Germany" and the many

thousands of Jews now being deported to them each week will soon be *de jure* as well as *de facto* deprived of their citizenship and their worldly goods.

Now free to rob and pillage their own citizens, the Nazis did just that—but not only in Germany. They also gathered the possessions of their new, involuntary subjects in France, Belgium, Austria, Poland, Hungary and Holland. They didn't demonstrate much patience either: in France, for example, the decree of General Keitel to render assistance in the confiscation of "ownerless Jewish possessions" was published on September 17, 1940, not even 5 months after the victorious German forces had marched into Paris.

The German efforts were successful—they found gold bars, jewelry and priceless works of art. No effort was spared in this exercise. The property was sold, exchanged, melted down or, if deemed appropriate, shipped to either Hitler's art museum in Linz, Austria, or the palace of Hermann Goering (head of the Luftwaffe) at Carinhall (to be donated for the greater good of the Reich and the *Volk* on his 60th birthday). The gold bars, together with jewelry and gold bullion, would eventually make its way to the *Reichsbank* to increase the gold reserves of Germany.

The story of the theft of personal ("non-monetary") gold by the Nazis is an ongoing one. To this day, there are various organizations and associations trying to ensure that reparations are paid to representatives of those from whom so much was taken. Although there is no way to calculate how much money was taken—most of it was melted down and used to mint new bullion—estimates put the figure at anywhere between $50 million and $300 million.

The Swiss Bank Accounts

Those individuals with enough foresight to realize the impending risk to their fortunes put their hard assets into Swiss bank accounts. While this was definitely the best thing for them to have done at the time, after the end of the war, only a very small percentage came to collect their assets. The reason for this, sadly, was that the majority of them had died in Nazi concentration camps.

The figures for this loss are, again, very difficult to calculate, particularly due to the stringent banking secrecy laws for which Switzerland is famous. In this particular instance, however, a number of circumstances conspired against the heirs to these hidden fortunes, not the least of which was the beginning of the Cold War. Eventually, however, in 1995, some progress was made on this issue. The Swiss Bankers Association met with the World Jewish Congress to resolve the issue.

That meeting eventually started a swarm of activity on the issue that led to a massive U.S. government investigation on the question of Nazi gold, and the forming of a number of independent commissions and bodies throughout Europe. These were dedicated to ensuring that, as far as possible, the victims of these Nazi financial crimes were satisfied in one way or another. These attempts go on to this day.

Germany's Trading Partners

Throughout the war, Germany was reliant on certain other nations to provide it with necessary goods and services. Sweden, Turkey and Argentina all gave considerably to the Nazi cause and, in addition to Spain and Portugal, were crucial to Hitler's continued waging of the war. (As an example, Portugal, one of the major beneficiaries of German gold during the war, received approximately 130 tons of it—worth at least $2 billion today.) While some (Sweden and Switzerland) maintained neutrality throughout the war, they continued to use their economic positions to profit from Hitler's war-mongering.

Turkey joined the Allies shortly before the end of the war, but Argentina, with its pro-Nazi regime, continued to assist Nazis long after the end of the war. Many Nazi war criminals escaped prosecution at Nuremburg through the secret underground that it created.

By, 1943, however, Germany's currency (the Mark) was no longer desirable to these nations or their industries and a viable alternative was sought. The simplest method was for Germany to use its gold reserves, and those of the countries that were now under its control. This additional pressure to make payment in gold, as opposed to using the German Mark, surely accelerated the process by which the central banks of Europe were summarily emptied of their gold holdings. As the need for gold kept growing, it was not long before the mouths of prisoners of Germany—especially in the death camps—were plundered for their gold teeth.

The Swiss Situation

Perhaps the most notorious neutral country during the war was Switzerland. Renowned for its resolute refusal to take sides in almost any conflict, the Swiss, next to the United States, were one of the few countries to make a profit out of the war, and hardly suffered any losses at all. As the world's financial and banking center, Switzerland continued to trade with Allied and Axis powers alike throughout the war.

More sinister than its continued provision of banking services to Nazi

Germany (which were given in spite of numerous requests by the Allies to stop) was Switzerland's provision of ammunition and other raw materials. This continued until the end of the war.

In general, however, nobody had really sought to place any degree of blame on Switzerland for the atrocities leveled on Europe by Hitler's Reich until recently. In 1996, a researcher using the United States' National Archive Records unearthed some documents, telling a rather disturbing story of complicity on the part of the Swiss government and the leading bankers of the nation.

Since the level of German gold reserves at the start of WWII were widely known, it can be inferred that the Swiss, who made over $2 billion worth of transfers for Nazi Germany during the years 1938-1945, must have known that the gold with which it was being paid could only have been acquired from the central banks of the countries which Germany had overrun. As such, Switzerland, some have said, assisted in the greatest money laundering operation in all history.

Switzerland was also the home to numerous secret individual accounts—accounts that, for example, would not need to be reported by Jews that were being deported from Germany. Unfortunately, as we all know, over 6 million of these people were never to return, and thus never able to recover money and gold (and other items) that they had deposited in their secret accounts.

In its widely publicized Eizenstat Report, the United States detailed the level of Swiss complicity in the looting of the central banks, and went so far as to say that the assistance of the Swiss, more so than any other neutral nation, enabled Germany to continue to wage war long after it would have otherwise been capable.

The report also detailed the non-cooperation (and, in some cases, failure to comply with early agreements) of the Swiss in making contributions/repatriations to those that could not recover their assets from their accounts.

In reaction to this report, and to economic pressure, the Swiss government has, in the past 7 years, initiated commissions to investigate these accusations and made further attempts to redress these matters.

The Hideout in Merkers

The majority of the stolen gold was kept by the Nazis at the *Reichsbank* in Berlin until 1943, when a portion of it was moved to various branch offices in order to reduce the risk of loss due to Allied bombing. As 1944 drew to a close and Allied bombers pounded Berlin, the majority of the gold reserve was moved out of the city.

On February 3, 1945, the Allies dropped over 2,200 tons of bombs on Berlin. The *Reichsbank* was nearly destroyed, along with its printing press. Due largely to this particularly close call, the Germans decided that approximately $238 million of gold reserves should be moved out of the city to a mine in the small town of Merkers. With legendary German organization, the gold (including gold brought back from the branch offices) was transported the two hundred miles to the mine.

Simultaneously, the majority of the remaining bank personnel were sent to the nearby town of Erfurt, together with a large amount of currency. The *Reichsbank*'s affairs were operated from there for the remainder of the war.

Bundled into the Merkers gold was also gold jewelry, cigarette cases, diamonds, false teeth, watches, gold and silver bullion and bars—valuables that had once belonged to prisoners in the concentration camps scattered throughout the east. It had been sitting in the Berlin bank until January of 1945, at which time it was determined that it needed protection; so off to Merkers it went.

In March of 1945, 45 cases of art works were also sent to the area for safekeeping. The mine at Merkers had become the new Nazi vault. By April it was decided to move everything back to Berlin, but the Allied forces were advancing so fast that it was impossible; the gold, the jewelry, the currency and the art all stayed there. A mere two months after the treasures had been put in the mine for safekeeping, the mine would fall into Allied control.

Soon after arriving in Merkers, the American troops started hearing that there was something valuable being kept in the various mines in the area. On the morning of April 7th, three soldiers descended the 2,100 feet to the bottom of the mine. Right outside the elevator they found 550 bags of German currency, but the main vault inside the mine was blocked by a brick wall. The soldiers summoned demolitions experts, and the next morning the wall was destroyed.

Over the next few days, an inventory of the treasure was made. This was what was left of the German pillage of the European central banks. According to the records, the following is what greeted those soldiers on April 8, 1945:

- 8,307 gold bars
- 55 boxes of gold bullion
- 1,300 bags of gold Reichsmarks, British gold pounds, and French gold francs

- 711 bags of American twenty dollar gold pieces
- 1,763 bags of gold and silver coins
- 80 bags of foreign currency
- 9 bags of valuable coins
- 2.76 billion Reichsmarks
- 20 silver bars
- 40 bags of silver bars
- 63 boxes and 55 bags of silver plate
- 6 platinum bars

The gold alone weighed over one hundred tons. At current ($400) prices, that's equal to approximately $1.4 billion worth of gold. Indeed, as remarked to General Patton at the time, if the old free-booting days were in operation Patton would have been one of the richest men in the world. Fortunately, however, it was to stop such activity that the Allies had fought Hitler in the first place. The whole experience had quite an effect on everyone—a room full of treasure, suitcases overflowing with jewelry, gold and silver bars and coins packed solid. That same evening President Roosevelt died, no doubt adding to the surreal feeling of the morning mine inspection.

By the time a more accurate calculation of the value of the Merkers treasure had been completed, it was determined to be over $520 million—more than $5 billion in modern terms. Now the Allies had to decide what to do with it.

The Aftermath

The efforts to restore the lost gold to its rightful owners have been extensive but not entirely successful. While serious impetus was given to these efforts directly after the war, the advent of the Cold War soon brought new pressures to bear on the governments attempting to make good the losses suffered under Nazi Germany.

Specifically, the Allied attempts to recover assets from neutrals, such as Switzerland, Sweden, Portugal and Argentina, were met with strong resistance initially. They were then superseded by the need to rebuild Europe, and then by Cold War imperatives and the need to strengthen the economy of the new West Germany.

The Allies gave $25 million to the International Refugee Organization for "non-repatriable persons," which included both Jewish and other survivors of Nazi atrocities. The Allies then had to decide what to do with the 337 metric tons of gold found in Germany.

The Tripartite Commission for the Restitution of Monetary Gold ("TGC") was established in September 1946 and was charged with the responsibility of recovering all monetary gold looted by Germany and distributing it among claimant countries. In addition to the gold uncovered by the Allied forces at Merkers and elsewhere, some gold was transferred to the TGC by neutral countries, although there are still allegations that some portions of the agreements reached between the relevant countries were never fulfilled. Nevertheless, the Swiss transferred 250 million Swiss francs to the TGC.

It should be repeated, however, that the Swiss have, during the last 20 years, made efforts to determine if there has been wrongdoing on their part and have made payments where they have found this to be the case. And there have been substantial grassroots movements that have contributed to the victims of Nazi Germany.

The TGC, by 1997, had distributed over 97% of the gold that it held. The balance was worth a little under $70 million at that time and a large portion of it was donated to the World Jewish Congress for the benefit of personal victims. The TGC shut itself down in September, 1998, announcing that after 50 years of operation, its task had been completed.

Japan's Plunder of Asian Gold

Nazi Germany was not alone in ransacking the banks of the countries that fell before it. Japan, which had been at war for most of the 1930s, couldn't resist the temptation either. Brimming with militaristic sentiment and national pride, the Japanese trained its people for war from the moment they were old enough to stand. It was only natural that this small, mountainous country would seek resources from outside its own border. And if it could gain access to minerals, farms and oil reserves, then why not also take command of the monetary wealth that could be found at the same time…

Although it is still a contentious issue as to the extent to which Japan's emperor condoned these thefts, there is little to disprove the belief that it took place.

Before World War II had even begun, the Japanese had stolen over 4,000 tons of gold from Nanking, then the capital of China.

Like the Germans, the Japanese had an extensive organization in place to ensure the efficient and effective removal of this gold. Allegedly named "Golden Lily" (*kin no yuri*), their prowess grew along with the Japanese military campaign. They followed the Imperial Army throughout southeast Asia, stealing golden Buddhas from temples and robbing the individuals that frequented them.

Once the gold had been collected it was melted into ingots in Malaya and marked with its purity and weight, then shipped back to Japan. Eventually, however, the center for the Golden Lily operation moved to Manila in the Philippines.

The Americans in the Pacific had an experience similar to their colleagues in Merkers when, in the Philippines, they managed to extract some information that led them to the Golden Lily sites hidden in caverns in the valleys of Manila. It is rumored that the gold that was collected during the first few years after the war was kept for the benefit of the fledgling CIA and was known as the "M-Fund." Allegedly, it has been used for everything from financing the Nicaraguan counter-revolutionaries to re-arming the Japanese army after the outbreak of the Korean War.

Ferdinand Marcos (one-time leader of the Philippines) conducted his own, rather successful, search for remaining gold in the hills of Manila and in the seas nearby. He is known to have used particularly unsavory methods of locating the gold but managed to locate at least $14 billion worth of gold (under current valuations).

Unlike the Nazi gold, which has been distributed as fairly as could be determined, the gold stolen by the Japanese appears to have been simply stolen again—by the CIA and Ferdinand Marcos.

Epilogue

On April 9, 1940 the *Wehrmacht* marched on Norway. While the soldiers and armored vehicles made their way into the heart of Oslo and the *Luftwaffe* landed in the city's airport, Norway's bank officials loaded 50 tons of gold into a convoy of trucks that escaped from the city just in time.

The German forces soon discovered that the gold which they had been hoping to find had eluded them. Planes were dispatched and the convoy was discovered and relentlessly bombed. Some of the gold was lost, but the rest joined the Norwegian royal family for a trip across the North Sea to the safety of the United Kingdom.

This gold survives until this very day—among the few European treasures that escaped Hitler's hands and the *Reichsbank's* melting pot. It stands today as a testimonial to the insatiable greed of the Nazis—and all of mankind—for gold.

CHAPTER 4

Gold and Economics in the 19th and 20th Centuries

During the 19th and 20th centuries, humanity began to experience an acceleration in progress. Industrialization and revolutionary technologies altered the face of business and the quality of life. At the same time, new experiments in government were bringing greater freedom and progress to some areas, and new forms of repression to others.

By the 1880s, nearly all of the world's major currencies were based on gold. Britain, France, a new Germany, and a still-young USA all adopted the gold standard. This shared basis for money and a wide freedom of movement between countries created a global community that in some ways we have yet to duplicate.

The 1800s were a time of relative stability. There were wars and recessions, but nothing like the turmoil that the world would witness in the 1900s. Even so, the 1800s did see several significant events worthy of discussion. The bearing of gold on these events varied—sometimes it was a key player, sometimes its hand in events was more subtle, and sometimes it was a lack of gold that shaped the way things unfolded.

Gold's influence on the 1800s actually began during the previous century. One of the first prominent persons to understand the importance of a fixed price for gold and its link to a currency was Sir Isaac Newton, the famous mathematician, scientist and natural philosopher. Late in his life he became an influential politician: the Warden of the Mint of England in 1696, and the Master of the Mint by 1699. Newton threw himself into his work and learned everything he could about economics, trade, currency systems, and the mint's operations.

The mint faced a problem. The day-to-day coinage of England was made of silver, but England also minted a gold coin called the guinea. By Newton's time, gold's value had begun to rise. A profitable trade had emerged: exchanging guineas for silver coins, and sending the silver to India. There, a silver shortage had developed and a guinea could be purchased for only two-thirds the silver required in England—and the cycle would then repeat. Thus, there was a constant loss of the mint's silver coins, since they were always being sent to Asia.

To avoid these problems Newton wrote a famous paper that put values on various weights of gold and silver in different countries and also assigned weights and values on its own currency. In 1717, based on Newton's studies and advice, the treasury established a fixed price of gold at 3.89 pounds sterling per standard ounce. That fixed rate remained unchanged for more than 200 years, and set the stage for what would follow.

The 1800s began on the heels of a French hyperinflation brought on by the revolution of 1789. Shortly after the revolution the French government issued a new currency: the assignat. They also passed a resolution that only 1.2 billion assignats would be printed.

It took the French Assembly only nine months to ignore their own edict and print more of the currency. By 1793, workers' wages had fallen too far behind the rising prices brought on by the government's loose printing of money, and rioting ensued. Rather than correct the root cause of the problems, the French Assembly passed laws capping prices and enacted stiff penalties for anyone caught refusing to accept assignats as payment for goods or debts.

The year 1795 began with fewer than 10 billion assignats in circulation and ended with 40 billion—and the public breaking of the printing plates for the assignats.

The French Assembly then repeated its mistakes with a second currency called the mandat. In just two years, the mandat's printing plates were also publicly destroyed.

In 1797, France began using a new currency based on gold and silver coins—a decision that began a widespread return to monetary standards based on gold and silver, though the French didn't adopt an official gold standard for some time.

Other events also occurred that began to pave the way for a worldwide gold standard.

Continued Expansion in the 1800s

As we saw in Chapter Two, the original thrust of colonialism from Western Europe into the western hemisphere was driven by a desire for gold. The Spanish explorers' discovery of gold on the American continents sparked the interest of France and England, and the race was on to establish a presence in the New World. The rush of colonialism changed the face of the globe forever.

By the 1800s, the heat of colonization had cooled, but the European powers continued to expand for other economic reasons. France and Britain both established new colonies in Asia, Africa, and the South Pacific. Portugal, Germany, Italy, Spain, the Netherlands, Belgium, and the United States also expanded their territories. The original motivator—gold—was not as much of a factor in these new acquisitions. Instead, the colonizing nations sought to build their trade networks and to increase their natural resources. But continued colonization meant that power was further solidified in the hands of a few, more technologically advanced nations, and the economies of the world became closely linked. This latter consequence made it more desirable for the world's nations to make the shift to a gold standard.

The Rise of the Gold Standard

What is a "gold standard"? There are different variations, but nations which adopt a gold standard "peg" their currency to gold in one form or another. The simplest way is to make a certain amount of currency exchangeable for a defined amount of gold. For example, the United States pegged the dollar so that one dollar was defined as 23.22 grains of fine gold. So one full troy ounce of gold equaled $20.67. (The $20 gold coins of the period contain a little less than one full ounce of gold.)

Why does a gold standard promote stability? A currency pegged to gold cannot be inflated, since more currency can only be printed as the country's gold reserves grow. This creates both internal and international stability.

Internally, the economy is much less volatile, since the currency itself can't be devalued by government printing of large numbers of unbacked notes. Internationally, trade is promoted when the trading nations are on gold standards. When all currencies are equivalent to gold, there are no fluctuations in currency values, and no problems converting one currency to another. More importantly, the flow of gold in and out of a country corrects trade imbalances. When a nation consumes more than it produces, more gold flows out than comes in, and a scarcity of money begins to occur. This will cause prices to rise for citizens of that nation. This discourages consumption and encourages production, which corrects the problem.

Gold critics sometimes claim that economies can't expand if the currency is anchored to gold. This is untrue; the gold supply itself expands every year as more metal is mined, so a currency pegged to gold can expand along with it. This rate of expansion is lower than modern politicians would like, but even modern economists agree that economies which expand too quickly will pay for it later in a recession. A gold standard allows for moderate, sustainable economic growth.

Realizing these advantages, in the 1800s nations began to adopt the gold standard. Aside from France's decision to return to the use of precious metals to mint coins, the other important currencies of the 1800s were the British pound, the American dollar, the German mark, the Australian dollar, and the Canadian dollar. Britain officially adopted the gold standard for the pound in 1821 and it was the first nation to officially do so.

Australia and Canada followed Britain's lead in 1852 and 1853 respectively. By 1879, France, which still used precious-metal coins, but also used banknotes not backed by gold, had officially followed suit along with Belgium, the Netherlands, Switzerland, Scandinavia, Germany, and the United States. Before long, nearly the entire globe was on a gold standard, smoothing international trade and providing much in the way of economic stability.

Other Important Events in the 19th Century

American expansion and gold

In addition to worldwide colonization, the United States continued to expand internally as well, and the push for this expansion was driven and shaped by gold discoveries.

Rather than a continuous push from the east coast out to the west with settlements cropping up along the way, the settlement of the United States was directed by gold discoveries. The most well-known—the California Gold Rush, which is discussed in depth in its own chapter—brought settlers from one coast to the other, leaving very few population centers in between. Settlement of the center lands was done in spurts largely following on the heels of new gold discoveries. This left vast spaces of wilderness, unpopulated and mostly untouched, in between settlements.

The discovery of new gold mines, as well as other types of mines, led to another interesting phenomenon: the advent of localized mining stock exchanges. Many mining towns experienced a boom, and in order to draw more capital and more people, local stock exchanges were opened—often dozens in a decade, in all different regions of the country and catering to different needs and interests. The popularity of these kinds of stock exchanges created two effects: for the first time, average people began to participate in speculative investing on a large scale, and the shape of the modern day corporation began to take form.

The Civil War and gold

The most expensive event any country can endure is a war. While there is a commonly held view that war is good for business and good for an economy, wars cost governments an enormous amount of money and use much in the way of resources—all without producing much of anything. Which is not to say that some wars are not necessary—just that they are expensive.

The most costly war of the 19th century in money and in lives was the Civil War in the United States. While there were ideological motivations behind the war, another point of the war was economic. The southern states felt exploited by the northern states and made the decision to secede and found their own nation—a decision that the North didn't accept.

In the war, the North held the advantage. Far more industrialized, the North was better able to produce the machinery of war. It was also better equipped financially. The North continued to hold states like California, a major producer of gold at the time. During the course of the Civil War, California and Nevada alone mined over $173 million worth of gold, much of which went to fuel the North's war effort.

Conversely, the South found it impossible to collect the money it needed to fund its war effort. Taxes covered only 5% of its expenditures. Attempts to borrow from its citizens in the form of bonds failed because the wealth holders of the South were mostly plantation owners, and the war interrupted their crop and disrupted their cash flow. Finally the South turned to printing more money and even declared counterfeit dollars valid (printing supplies were short and it couldn't print what it needed fast enough).

The South experienced steep inflation as a result of this monetary policy. The cost of living was 92 times higher at the end of the war than it was at the beginning. And it proved only a matter of time before the South went broke, and was forced to return to the Union.

The Boer War over gold

While the influence of gold had a notable and important effect on the outcome of the Civil War, gold itself was the cause of the Boer War in southern Africa.

The Dutch colony of Transvaal in southern Africa proved to be the home of one of the largest gold fields on the globe. The gold was discovered in 1885, and resulted in an influx of unwelcome British settlers hoping to make their fortunes through mining. The Dutch colonists did not afford the British newcomers the same rights as Dutch citizens. Eventually resentments between the Dutch and British grew heated, and war broke out.

The Boer War lasted for three years, from 1899 to 1902. The British achieved a shallow victory. They relieved the Dutch of their gold-rich land but compensated them 3 million pounds and promised they would eventually have their own government. By 1910, the Union of South Africa was established.

The Gold Standard Ends: Economic Turmoil in the 20th Century

While the 1800s saw many major changes, some brought about through upheaval and tragedy, it also ended during a period of unprecedented global stability, largely due to a nearly universal gold standard. This was not to last, however. The 1900s would see much in the way of prosperity and technological advancement, but it would also prove to be a century of widespread turmoil and instability.

The Aftermath of World War I

At the turn of the 20th century, power had been consolidated into the British, French, German, American, and Russian nations. The global stability that was accomplished in the late 1800s held fast, until it was shattered by the First World War.

This war had far reaching implications.

All but the United States went off the gold standard when faced with the mounting expenses of the war. Raising taxes was too unpopular, and each government found their gold reserves could not meet their needs. From the governmental point of view, inflation became desirable.

Once the war had ended, it became clear that because of the enormous debts that countries had incurred and because of currency devaluations, the gold standard could not be restored right away. And not long after, many of the countries that had fought in the war experienced a new horror: hyperinflation.

Hyperinflation

What is hyperinflation? The term itself is not well defined. Some use a measure of consumer price index, and define hyperinflation as a 100% increase in CPI over a year's time. But the most commonly held definition is more vague—a steep rise in inflation over a short period of time. We would add to that definition that a hyperinflationary period often ends with the destruction of the currency in question.

Germany's post World War I hyperinflation is probably the best-known example.

At the beginning of 1919, 170 German marks would buy an ounce of gold. By November of 1923, a single ounce of gold cost 87 **trillion** marks.

The German hyperinflation was a direct result of the war. The war left Germany politically unstable. Many of its former trading partners now viewed Germany in a suspicious light as they had become enemies during the war. Political difficulties, problems resuming previous trade patterns, rising costs of food and raw materials, and the harsh reparations leveled against it, all meant that Germany was headed for disaster.

The war reparations were the main issue. The Allied nations settled on a sum of 132 billion marks to be paid over time, a number that Germany insisted its gross national product could not support. The burden of reparations accelerated the printing of money.

The German experience in 1919-1923 was a classic example of hyperinflation. It started slowly, but accelerated throughout. As with other hyperinflations, toward the end the currency was collapsing so quickly that the government couldn't print notes quickly enough. By the time the notes were distributed, they were almost worthless. Factory workers were paid twice a day so their wives could rush to the shops to buy something—anything!—while the notes still had a little purchasing power.

It ended at last with the cessation of the currency. The German mark was replaced with the Roggenmark—citizens were to turn in 1 trillion marks of the dead currency to receive one Roggenmark. Unfortunately, during the hyperinflation, because the money's value was fading so fast, people spent it as quickly as they could. Very few had any to turn in and so by and large, the general populace was as bad off as before.

Other Hyperinflations

Germany was not the only nation to experience a post-World War I hyperinflation. Armenia, Austria, Azerbaijan, Danzig, Georgia, Hungary, Poland, and Russia all suffered through a period of hyperinflation after the war.

The Austro-Hungarian empire broke apart after World War I. The Austrian capital lost most of its industrial and agricultural resources as a result. With no industry to support it, the capital soon found itself in desperate straits. To provide relief to its people, the Austrian government began printing money. By 1922, the Austrian krone had depreciated to

1/2000th of its pre-war value. In 1924, the krone was replaced with the schilling, an issue of the Austrian National Bank.

Severed from the Austrian empire, Hungary was also thrown into a period of chaos. Both Soviet forces and Allied forces wrestled for the nation. Industry stalled out, the government was bankrupt, and inflation began to soar. In 1921, a man named Count Bethlan assumed power. He acquiesced to all Allied demands and was able to procure a $50 million loan for the reconstruction of Hungary.

Inflation was stopped due to his shrewd actions, and eventually Hungary established a new gold-backed currency, the Pengo.

Meanwhile, Russia's shift from monarchy to communist nation was certainly not a smooth one. And the new Russian leaders had a liberal view of the printing press that didn't help matters. Russian hyperinflation reached the point where ruble notes were printed off in denominations of billions. Armenia, Azerbaijan, and Georgia were all affected by Russia's monetary problems.

World War I reshaped much of Europe. It also altered the global balance of power. Before the war, there was little fluctuation in exchange rates—international trade was a balanced affair. After the war, the economic make-up of Europe was chaotic and in a constant state of flux.

Nations also saw a new need for self-sufficiency. Where they once had been willing to specialize in what they produced and depend on imports from other nations for any unmet necessities, after the war countries felt they had to be able to meet their own needs.

Getting Back on Track Between the Wars: a Dismal Failure

Except for the United States, every major nation went off the gold standard during World War I.

Afterwards, the European nations struggled to return to it, but they were unwilling to use the new inflated values of their currencies to adopt an adjusted gold standard. They wanted things to be as they had been before, which proved impossible.

At best, Britain was able to implement a partial gold standard, where it paid international debts in gold, but made it impractical for its citizens to trade in their pounds for gold. These partial gold standards did not control inflation well and created further imbalances among the nations of Europe as they tried to pay off their debts to each other. In 1931, Britain went off the gold standard completely, and the rest of Europe followed. Such a move was usually devastating to the country's citizens;

for example, within three months after Britain dropped the standard, the British pound lost a third of its value.

Even the United States' commitment to gold wasn't to last. Rumors began to swirl that the U.S. would also drop the gold standard. To protect themselves, people began to withdraw their savings from banks and convert their dollars into gold (as was their right). This caused the collapse of thousands of banks, which had overextended themselves. Rather than standing firm on the gold standard and fix the underlying economic problems, President Roosevelt's solution was first to confiscate the people's gold, and then take the nation off the gold standard anyway.

Under Roosevelt's new monetary laws, in 1933 it became a criminal offense for an individual to possess gold bullion or coins. In 1934, the U.S. Treasury collected all the gold held by the Federal Reserve banks. For foreign exchange transactions, the dollar was devalued. While it previously took $20.67 to buy an ounce of fine gold, it now required $35.

The chaotic monetary policies in Europe set the stage for the beginning of the Second World War. The world came out of World War II better than it did World War I. It seems that many lessons about nation building and establishing new currencies were learned from the first war.

After the war, the German currency faltered, but Germany was able to successfully establish a new currency, the Deutschmark. Having learned that simply exchanging the old currency for the new at a set rate was insufficient to jumpstart an economy, this time around they issued everyone 60 Deutschmarks, as well as exchanging old currency for new.

Though instances of hyperinflation were fewer, they did occur in China, Greece, Hungary, Indonesia, and Romania. China's and Hungary's situations were the most interesting.

China's yuan had been based on silver and was quite stable. When, in 1934, the U.S. government began purchasing silver to add credibility to its own silver certificates—this in response to the abandonment of its gold standard—China sold too much of its silver reserves to the U.S. and deflated its currency in the process. In response to the currency crisis, China nationalized its banks and required its citizens to turn in any silver they had in exchange for banknotes. World War II depleted China's resources and the nationalized bank turned to printing money to fund its fight against communism, driving China into a period of hyperinflation that eventually ended with the Boxer Rebellion and the shift of China to a communist nation.

Hungary's gold-backed pengo proved a stable currency during the period between the wars. However, Hungary faltered during WWII and cast its lot in with Nazi Germany. After Germany's failed attempt to topple Russia, the German government stole Hungary's gold reserves during its retreat from Russia back to Germany. Depleted of resources, Hungary experienced a record-breaking hyperinflation, going from 3.5 pengoes to a dollar in 1941 to the unpronounceable figure of 4,690 trillion trillion pengoes to a dollar in 1946.

But overall, there was less turmoil than previously. The Allied nations realized that punitive post-WWI reparations had not prevented a second major conflict, and took pains to rebuild the aggressive nations so that they might not grow aggressive again. In this they were very successful—Germany and Japan remained militarily weak after World War II, but their economies flourished.

Finally, the major global powers sought to return the world to some sensible international monetary standard and restore global trade patterns. The solution they came up with is often referred to as "Bretton Woods," named after Bretton Woods, New Hampshire, where the international monetary conference was held in 1944.

The Bretton Woods system attempted to bring some stability to the hodge-podge gold standard used between the two wars. Instead of many key currencies, there would be only one—the U.S. dollar. The dollar would be valued at $35 per ounce of gold and would be used as the standard to settle foreign debts and conduct foreign trade. It would continue to be illegal for individuals to possess gold—only governments and their central banks would be issued gold in exchange for dollars. This new system offered some stability, but it did not check inflation, it only slowed it.

(The Bretton Woods system is vital to understanding our modern global economy, and why gold is poised to move so powerfully today. We'll discuss it further in Chapters Eight and Nine.)

The Rise of Economic Instability

The seeming stability after the Second World War was short-lived, and the second half of the 20th century saw an unprecedented number of hyperinflations. In 1971, the decisions at Bretton Woods were abandoned and the world turned to a complete fiat system of money, the implications of which are discussed in detail elsewhere in this book.

One of the results, though, was that hyperinflations became more commonplace. In the last quarter of the 20th century, Angola, Argentina, Belarus, Bolivia, Bosnia, Brazil, Croatia, Georgia, Nicaragua, Peru,

Poland, Turkey, Ukraine, Yugoslavia, and Zaire have all suffered through periods of hyperinflation.

Bolivia saw a 24,000% increase in consumer price index in a single year in the mid-eighties. Argentina's hyperinflation and depression were so severe that people starved to death in the streets. Yugoslavia also had a particularly brutal hyperinflation.

The rise of unaccountable monetary policies has brought tremendous suffering to many, many people across many different nations.

So what is to be learned from all this? Despite our progress in industry and technology—which has been impressive—gold continues to be a shaping force for the world and its economies. The gold standard, when widely used, created stability and trust between nations. It improved trade and increased globalization. It solved far more problems than it created. Without a gold standard, governments become solely responsible for the value of the currencies that their people depend on—and too often governments abuse that power, leaving people bereft of their savings and without a meaningful standard of trade.

And what will happen now?

Ahh—we shall see shortly. But first, a little more colorful history, and primer on how gold is recovered and how much gold remains to be found.

First, let's look at the greatest migration the world has ever seen—fueled by the thirst for gold...

CHAPTER 5

The California Gold Rush

Sam Brannan had heard the rumors of gold at Sutter's place. He ran the only general store within miles of John Sutter's little empire, so how could he not? Still, he was skeptical.

But then the rumors became commonplace, and a few miners with hopes of big fortunes began to straggle in. And occasionally, a customer would pay with tiny gold pebbles.

Sam had no desire to make his fortune mining. It was uncertain work and dirty, but he did see a way to make his fortune from gold. A plan occurred to him. Quietly he began buying up all the mining tools in the region. He bought every pick, shovel, and mining pan that he could find. Finally he was ready.

For his plan to work, he needed to generate some far-reaching excitement. He needed to confirm the rumors of gold once and for all and do it in a way that would really get people talking.

On May 12, 1848, he took some of the gold he had collected over the last weeks, mostly from his customers, put it into a vial, and ran through the streets of San Francisco, showing the gold to startled onlookers and shouting, "Gold! Gold! There's gold in the American River!"

Sam Brannan's little publicity stunt was the catalyst for the Gold Rush, and the largest global human migration since the ice age began. And while the Gold Rush really didn't warrant that name until after Sam Brannan's publicity stunt, the real event that started it all took place nearly five months earlier, on January 24, 1848.

Gold in the American River

John Sutter had been born in Switzerland. He'd traveled quite a bit, but upon arriving in California in 1839, he immediately recognized its agricultural potential and determined to stay there and make his fortune.

Sutter arrived in California at Yerba Buena, then an unremarkable town of just a few hundred settlers, but later to become San Francisco.

He made his way up what is now known as the Sacramento River and found a suitable place to settle. His holdings, commonly known as Sutter's Fort, would later grow to become Sacramento. Though he was local to Sacramento—and the events surrounding the Gold Rush did change Sacramento—the major impact of the Gold Rush was on San Francisco, as it was considered to be the gateway into gold country.

Within a few years of settling, Sutter had acquired 12,000 head of cattle, he had vast vegetable gardens and orchards, he employed hundreds of workers, and he had several small but growing industries to support his growing community.

One of these, a grain mill, needed more lumber to be complete. A sawmill was in order.

Sutter hired James Marshall to build him one on the American River. Work on the sawmill began, and soon afterward Marshall noticed small yellow pebbles in the water. He picked up a few and realized what he had found.

Marshall was a good man and loyal to his employer. He kept quiet and took the gold he had found to Sutter. The two men conducted some rudimentary tests and the results were enough to satisfactorily confirm what they both suspected—Marshall had found gold. They both agreed that until the sawmill was finished they didn't want rumors of gold to spread. But that presented a difficulty—Marshall was not the only man to have found gold…others among the workers had, too.

Sutter rode up to the work site the next day to ask the workers for just six weeks of silence. The workers agreed, but as Sutter rode home, he knew how unlikely it was that the rumors of gold would not spread.

The eventual flood of fortune-seekers into Sutter's lands ruined his dreams of an agricultural enterprise. His gardens were raided, his building stripped for resources. In his own words, "What a great misfortune was this sudden discovery of gold for me! It has just broken up and ruined my hard, restless, and industrious labors, connected with many dangers in my life, as I had many narrow escapes before I became properly established."

Still, the event begs the question: Why did the California gold rush become an event of such great magnitude?

Why California in 1849?

For all of history, the majority of cultures have been willing to migrate for gold. Gold brought Europeans to the Western hemisphere, it brought the British to South Africa, but never has a gold discovery so incited the world to action the way the Californian gold discovery did.

There were two primary reasons that California gold was able to take such a firm hold of the dreams of men—and sometimes women—the world round. The first was geological.

The gold discovered in California was placer gold—found on the surface and in the riverbeds. Easy to see and easy to get to. Not much work was required beyond locating it and then picking it up. Fortunes didn't come any easier.

The second reason for the magnitude of the California Gold Rush was government. The role that government played in the early part of the Gold Rush is both limited on one hand and pivotal on the other.

The Gold Rush really happened in two parts. In 1848, after Sam Brannan's brash, public announcement, people local to the region flooded to the area to find gold. Many of these did indeed strike it rich, quickly and without much toil.

But it wasn't until 1849 that a fever gripped the world, and thousands made their way to California to strike it rich. In 1848, nobody but the locals really believed the rumors of gold. They seemed too fantastic. Or rather they did until President Polk confirmed that the rumors were true, publicly stating, "The accounts of the abundance of gold in that territory are of such extraordinary character as would scarcely be believed were they not corroborated by authentic reports of officers in the public service."

Sam Brannan's announcement in the streets of California brought the locals running. But it took a presidential statement to have the same affect on the world.

Government confirmation got people moving. But government played a second, very different role in California.

California was not yet a state when gold was discovered. And this meant there was not yet clearly written (or enforced) property laws. For example, why exactly were people just allowed to invade Sutter's property in droves? It was because his claim to those lands was tenuous, at best. He hadn't really needed to prove a legal claim to the land as no one else seemed to want it.

After the discovery of gold, Sutter's property was a point of interest. He knew his lands were in danger and so he negotiated to buy it from the local Native Americans, but the documents that he drew up for this transaction did not hold much legal weight with the courts—property law was still flexible and the courts decided Native Americans did not have the right to transfer or grant ownership of land.

This lack of governmental structure in California created the real appeal. No established government meant little in the way of regulations on mining and who could claim what. The gold was literally lying on the ground for the taking, and it was on a first come, first served basis.

This promise of an easy fortune struck to the very heart of tens of thousands of dreamers and doers around the globe.

Where They Came From and What They Left Behind

The 49ers, as they came to be called (for the year, 1849), were nearly all men, and they came from all over. From the eastern seaboard, from the central states, from Chile, from Mexico, from Ireland and France and Germany, from Turkey and China. It seemed no one was immune to the hope of striking it rich, and striking it rich quickly.

Most left behind families. Just for a year, they planned. Surely that was all it would take to achieve their version of the American dream.

California and its gold held the promise that the 49ers could, within a year, have a fortune large enough to get them out from under a mortgage, pay off their debts, and free them from their jobs forever. This promise of freedom was a common motivator among the 49ers.

The dream was that they would make their way to California, pick up enough gold to last a lifetime, and speed home to their families. They planned for this journey never knowing that most of them would only find disappointment. And that so many difficulties lay before them.

First, there was the task of getting to California. At the time, it was remote. No railroad traveled there. The options were to spend months aboard a ship or months traveling over land. Both were perilous.

A ship journey around the tip of South America—preferred by those on the east coast—involved seasickness, rotten food, stale water, and incomprehensible boredom.

A ship journey to Panama and then over land to the Pacific shore to catch another ship could prove shorter, but involved the same problems and also increased the risk of malaria and cholera.

Traveling over land—the choice for those from the central states—involved months of walking and hunting and trading. These fortune-seekers were rarely experienced frontiersmen, making the journey all the more difficult. And then the last stretch of the journey was perilously dry...people commonly died of thirst.

But the lure of the gold in the west—and the promise of a better life that they took to their heads and hearts of a better life—persuaded most to press on.

Just Enough Gold to Keep Dreams Alive

If Sutter's land had been the only place where gold was found, the California Gold Rush might have fizzled and died before it became such an historical event. But miners spread out all through the Sierra Nevada mountain range and found gold nearly everywhere they went. The same year as the American River discovery, gold was found in the Feather and Trinity Rivers. It was found mixed in with quartz deposits in Mariposa County and in Grass Valley. In 1851, gold was found in Kern County. And in 1852, discoveries were made at American Hill near Nevada City and in Placer County. More discoveries were made in 1853 in Tuolomne County and in 1854 in Calaveras County.

New discoveries fueled the push, but before the arrival of the 49ers, most of the accessible gold—the surface placer deposits—had been exhausted at the hands of local and regional miners.

The 49ers dreams of easy money were quickly replaced with long days of difficult work just to recover a little gold. Many worked 10 hours a day sifting sediments from the cold river bottoms, hoping with each new day to find their fortune. For most, each new day meant eking just enough gold out of the river and ground to keep going.

A Strained Economy

The bits of gold that most 49ers did find might have been enough to achieve some of their dreams and return home richer than they had left, had the California economy not been so affected by the sudden influx of gold.

In the 1850s a **half billion** dollars worth of gold passed through San Francisco. This had no small effect.

The sudden massive addition of gold to the economy and the huge influx of people into the area caused definite problems.

Gold was plentiful, but goods were not. Food was scarce, too. So food and goods became expensive in terms of gold. What may have added up to a fortune over time, the miners were forced to spend just to get by in the tumultuous California economy.

The changes in the economy did present opportunities to those with the eyes to see them. The biggest fortunes were made not from mining gold—but from recognizing the changes gold was making in California and capitalizing on them.

Many men and women got rich simply by recognizing one of the many new needs faced by the economic and population boom and figuring a way to fill them.

Sam Brannan was by far the most successful entrepreneur of the Gold Rush. His plan at the outset of the rush worked like a charm. He marked up his mining supplies 7500% on what he'd paid for them, and within nine weeks he had made $36,000, not an insignificant sum for 1848.

He used this strategy over and over, creating a monopoly on a certain type of goods and then capitalizing on it. By the mid-1850s he was worth half a million dollars

Unfortunately, he eventually drank away most of his fortune and died a poor man.

There are many famous success stories from the Gold Rush. One man with a well-frequented dry goods store made a fortune selling sturdy, canvas pants. Today, Levi Strauss is still famous for its jeans. An experienced butcher came to California for gold, but ended up opening a meat packing company, and today Philip Armour's company is still one of the largest meat packing companies in existence. A wagon-maker founded the beginnings of Studebaker. Two friends provided reliable banking, transportation, and mail delivery services—and Wells-Fargo was born, and is currently a $28 billion company (in annual revenue) with assets of close to $400 billion dollars.

The Changes in San Francisco

The small town of San Francisco was forever changed by the gold rush. It became the cultural center of the world, practically overnight.

In 1848 San Francisco had only 800 citizens. By 1853, it had grown to 50,000—and that didn't include the tens of thousands of people that might pass through the city on any given day.

Its growing pains were sharp. Six times the city burned nearly entirely, and each time it was rebuilt. Crime rates skyrocketed, and the sudden coming together of people from so many different backgrounds and ethnicities created tension and fostered resentment.

But it also fueled the cultural growth of the town. San Francisco became a home to opera houses and theaters. It was the envy of many great cities across the world.

The mining community helped to spur on San Francisco's growth. Many of the miners found themselves far from home and lonely, unable to let go of the reasons they'd come, but unable to fulfill their goals. They were desperate for a sense of community, and an entire industry built up around their need for connection.

Unfortunately, this industry was destructive more often than not—drinking, gambling, and other vices served to give the miners a way to deal with their frustration, while also keeping them poor.

More and more men coming to mine for gold meant that claims became more competitive and bandits became a problem. Crime of all types was on the rise, and the California government was getting more and more organized to deal with the new problems.

At the onset of the gold rush, the area was a peaceful, small town. Within four short years, it was a booming metropolis where hangings were commonplace and any manner of entertainment—wholesome or otherwise—was always near at hand.

Striking It Rich or Going Home Empty-Handed

By 1852, California gold mines hit their peak production of $81 million annually. And some miners did walk away rich. But most made just enough to pay their living and mining expenses.

Some packed it up and went home, finally realizing that they weren't finding any gold and it wasn't worth the time away from their families.

But for many, the grip of gold fever held them fast, and they lived out their lives—often tragically short—chasing after a dream that just wouldn't seem to come true.

The fortunes made in the Gold Rush that carried any lasting story with them were made by entrepreneurs. One interesting note about entrepreneurship in California was that the Gold Rush created an economic climate that destroyed many class and gender barriers, creating opportunities for folks that wouldn't have had them elsewhere.

Women, in particular, were able to flourish in the new and strange economy. There was a need for services and nobody cared who provided them. Women made money washing clothes, baking pies, running boarding houses. Many women who never would have had the opportunity for financial independence in a different city or region, were able to make a fortune doing things that in a different time and different place would have just been expected of them.

The Demise of the Individual Miner

Eventually it became clear that the day of individual mining and fortunes had passed. The surface placer deposits weren't just picked over, they were exhausted—and no new strikes were forthcoming. There was still gold to be had, but not in any form that an individual could hope to mine profitably.

To get at the gold that was left would take cooperation.

At first, informal companies were formed. Men would work together to build a dam and reroute a river so they could get better access to the gold deposits.

But soon, even that was not enough. More capital and new techniques were required if the remaining gold was to be reached. Corporations formed to achieve this end. Miners weren't happy about becoming employees, but they saw it as the only way to continue to chase after their dreams of gold. In some cases, they had become accustomed to the mining life and corporations offered them the opportunity to continue on with what they knew.

Corporations could use methods individuals could not. They sunk deep shaft mines. They brought in machines to scoop great amounts of soil and rock out of the riverbeds to sift through for gold deposits. They used hydraulic jets, powerful enough to kill a man at 200 feet, to tear into the riverbanks.

Hydraulic mining was effective, but it proved devastating to California's river systems, which still have not fully recovered. Thirty years after hydraulic mining's rise, the California government passed regulations preventing its further use.

The End of the Gold Rush...Its Lasting Impact

The Gold Rush has had a lasting effect on California. It brought scores of adventurous settlers and sharpened the eye of entrepreneurs to opportunity. It created a cultural center forced to integrate a large number of ethnicities.

It hasn't always risen to the task of dealing with its complicated history and culture in the best way, but California and the people living there have been willing to take risks that others haven't, and it has become the center of many ground-breaking industries—most notably film and computer technology.

You cannot explore the Gold Rush without the incredible lure of gold becoming apparent. The promise of gold placed hopes and dreams within reach, and people crossed oceans and continents—and literally moved mountains—for their chance at the opportunities gold offered.

NOTE: The Comstock Lode, a famous discovery made in 1859, is situated on an eastern spur of the Sierras, extending into Nevada. Placers and veins similar to those of the Sierras are found also in Oregon and Washington. The Rocky Mountains and the outlying ranges, which were first prospected in the early 1860s, include an immense area of gold-bearing territory. Rich gravels have been worked at the following locales: near Leadville, Fairplay, and in San Miguel County, Colorado; near Helena and Butte, Montana; along the Snake and Salmon rivers, Idaho; near Deadwood, South Dakota; at Santa Fe, New Mexico; and in Alaska.

Placer deposits of gold were also discovered on the Yukon River in Canada and Alaska in 1869. The discovery in 1896 of a rich deposit in the Bonanza Creek, a headwater of the Klondike River, which in turn is a tributary of the Yukon, led to another gold rush. In 1910 discoveries were made on Bitter Creek, near Stewart, British Columbia. In 1911 gold was also discovered in Alaska, some 30 km (some 20 mi) from the Canadian boundary at the source of the Sixty Mile River, which rises in Alaska and flows into the Yukon River. Since 1911 the production of gold in Ontario, in the Porcupine and Kirkland lake districts, has gone ahead rapidly. Important discoveries of gold mixed with copper were made in northwestern Québec. The gold production of Australia has been famous since 1851; the chief centers of production are in Western Australia and Victoria.

Part II

Where Is Gold Found, How Is It Obtained, and How Much Is Left?

CHAPTER 6

Modern Gold Production

Modern gold production is mostly about getting gold out of the ground at an economically feasible cost so it can be used in coins, bullion, and jewelry. (There are more and more industrial uses for gold, but we'll get to that in a minute.)

While this type of gold production is not exciting to most people, it's a turn-on for others. Especially those involved in gold exploration and the producing process, and "gold bugs"—those devoted investors who champion gold for investment, personal, and political reasons.

A few other interested parties are investors in gold stocks, hard-rock miners with picks and shovels looking for new places to mine, and weekenders panning for their fortunes in what were once ancient alluvial fans that spread the gold over wide areas.

Getting Gold from the Ground to Your Wedding Band (Or Smart Card)

Gold is all around us. It's in the sea, it's in most rocks, it's in little specks of gold and small pieces (nuggets) in riverbeds, and it's in veins deep underground.

The problem is finding and producing gold economically. That sometimes depends on the quantity of gold in the ground, and its location. Gold may be everywhere—but it's not necessarily easily recovered.

For example, the gold ore may be accessible, but other factors can make the task economically unattractive. There may be very little gold in a specific gold claim area, and the effort required and cost of digging it up and separating it from other elements may be prohibitive. Or if there is a lot of gold, the estimated yield of gold per ton of earth and rock that must be moved, crushed, and treated is not sufficient within the gold claim area to justify the project. That may occur if the gold lies very deep within the earth.

To complicate it even more, for any location it may be economical at one time and not at another, depending on the current price for gold.

And if the price of gold is volatile, the producer has to guess—before beginning recovery—what the gold may sell for after it's recovered.

There's Nothing Romantic About the Recovery Process

History may tantalize us with stories of huge gold nuggets found in a stream or cave, or buried treasure stumbled upon. But most of the yellow stuff comes hard and at great expense to gold producers. While some miners still pan for gold in streams and find nuggets, most gold comes from hard rock, and in locations where most people would rather not spend a lot of time. Most gold claims are not Club Med vacation spots. Some are in remote, humid, bug-infested areas of Asia or South America, some are deep in mines in South Africa, and others are in desert locales.

Olmsted and Williams, in their book, *Chemistry: A Molecular Science* (2001, p. 935) describe how getting the final product, once you locate and extract it, can be a tough job. They point out that gold in quantity normally does not come ready made in huge nuggets. It comes with baggage, or waste. The industry's name for this waste—organic material, clay, and sand—is gangue (pronounced "gang").

If you want the gold, you must separate it from the gangue using methods that will ultimately pay for the costs of exploration, mining, and refining the gold, and still leave a profit.

The two common ways to separate the gold from the gangue are by using one of two methods that have a long history in gold mining. The methods are called flotation and leaching.

Flotation

Flotation is the physical method of separating the gold. At first glance, this appears counter-intuitive. How would you get gold to float? You can, in the right process. The crucial step is adding a gold-attracting element at the right time and in the right manner.

First, since most gold production involves getting tiny specks of gold from rocks, you must crush the ore to expose the bits of gold. Then you mix the crushed ore with water in a large container, or vessel. The mixture becomes a thick "slurry." You then transfer the slurry into a large container built for the flotation method.

The flotation part comes after you add oil and a surfactant to the slurry. A surfactant is a "surface-active substance," which works something like a detergent that divides dirt from clothing.

Olmsted and Williams (*Ibid.* p. 935) explain it this way:

> "The polar head groups of the surfactant coat the surface of the mineral particles, but the non-polar tails point outward, making the surfactant-coated mineral particles hydrophobic. Air is blown vigorously through the mixture, carrying the oil and the coated mineral to the surface, where they become trapped in the froth."

Because the gangue has a lower "affinity" for the surfactant, Olmsted and Williams report, the unwanted part of the ore "becomes 'wetted' by the water and sinks to the bottom of the flotation vessel."

You then skim the froth from the top of the vessel, and the gangue remains at the bottom. The gold is now fairly easily separated from the froth.

The other common method of recovering gold is called leaching.

Leaching

Leaching ore is the chemical technique of separating the gold from the hard rocks and the gangue. It uses a solution of sodium cyanide and relies on the solubility properties of the ore.

Here's how it works:

As in the flotation method, you must first crush the ore, which is then put into something called a "pad."

A "pad" is a pile of crushed ore stacked on a plastic liner that blocks the cyanide solution from leaching into the earth below. The liner and slope of the pad direct the fluid runoff into the processing system.

When the ore is on the pad, a wetting system (drip or sprinkler) delivers the cyanide solution to the ore. The solution covers and penetrates the pad of ore, dissolving the gold it touches, then the gold in solution flows into a plastic-lined pit or holding area. From there, the gold in solution is pumped through the recovery process.

The recovery process involves a chemical extraction method that "frees" the metals in the solution. Essentially, when you expose the cyanide solution carrying the gold to zinc dust, the zinc attracts the gold, pulling it out of the solution.

Pads of ore range from a few yards to several yards deep. They can be several yards wide to hundreds of feet or meters wide and just as long. In small operations, gold processors fill square and rectangular pads by dumping ore using front-loaders. In larger operations, huge dump trucks and earthmovers drive onto the pads and drop the ore, creating huge piles that grow deeper and deeper as truck drivers dump more loads.

Some gold producers use a machine known as a Merrill Crow unit to capture the gold. These tank-like units have several long tube-like bags, or sleeves, that hang in the tanks. The cyanide solution passes through the sleeves and out into the tanks. Zinc dust is placed in the Merrill Crow sleeves, and the gold attaches to the zinc dust, pulling the gold from the chemical mixture. The Merrill Crow then sends the cyanide solution on to a retaining pit. From there it goes back into the sprinkler or wetting system.

When the sleeves have filled, crewmen pull them and replace them with fresh ones. The wet zinc dust holding the gold goes into a container. Then the dust is heated, producing a blob of metal known as a "button" that contains gold and other minerals. Heating the button to the proper temperature divides the gold from the other minerals.

Again, all this production goes on in the context of the monetary value of a certain quantity of gold mined within a specific area or by a mining company during a defined period of time. When gold prices are low, many mines with high costs can't operate profitably, so they stop production. When gold prices go up again, these mines go back into production.

Current Rates of Gold Production

Gold production is most often reported in ounces and tons (or tonnes). A U.S. ton is 2,000 pounds. But much of the world uses the roughly equivalent (but slightly heavier) measurement of metric tonne. A metric tonne weighs 1.1 times a U. S. ton.

That may seem like a small difference, but do the math. A U.S. ton is 2,000 pounds, while a metric tonne is 2,204.6 pounds. At 14.6 troy ounces per pound that's a difference of over 2987 ounces. At $400 an ounce, that's $1,194,800. (Next time you're buying a ton of gold, make sure you get a metric tonne and not a U.S. ton.)

If you like, you can get a better understanding of gold production historically at the website: www.goldsheetlinks.com/production2.htm. The site includes historical data compiled by the U. S. Geological Survey (USGS), the national authority and clearing house for such information.

For example, according to the USGS, worldwide production of gold rose from almost 400 metric tonnes a year in 1900 to just less than 2,600 metric tonnes in 2000. So we're now producing 2,200 tonnes a year more than at the beginning of the last century.

Also, consider this: Seventy-five percent of all gold *ever produced in the world* has been produced since 1910. That's according to the Gold Sheet Mining Directory at www.goldsheetlinks.com.

What's more, the Gold Sheet states that 50 percent of all gold ever produced in the world was produced *since 1960*!

Gold production viewed in relation to other long-term social trends produces interesting numbers, too.

Gold production growth exceeded population growth for 100 years, from the 1840s to the 1940s. Then production growth declined after WWII for almost 50 years. But it picked up steam again in 1989.

Put in per capita terms, the long-term trend is about ¾ of an ounce of gold per person each year, reports the Goldsheet Mining Directory website.

Take a look at this chart of gold production since 1900:

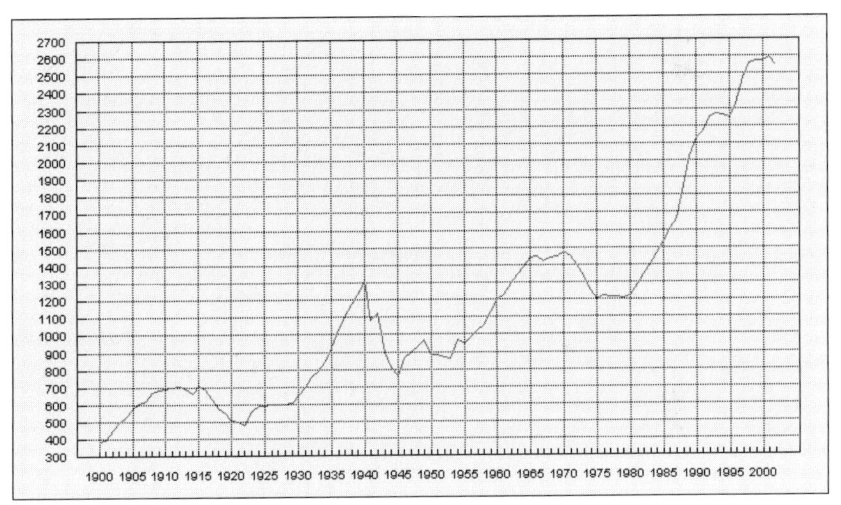

Figure 6.1: Gold Production Since 1900
(Source: www.goldsheetlinks.com)

You can see the upward movement in gold production, with only four "dips" that reflect both economic difficulties and war years. In fact, if you smooth the peaks and valleys in a gold production trend line, you find a strong upward movement that is steep since 1980, with only a blip in the 1990s when stocks were roaring and gold was languishing toward modern-day new lows.

Looking at the annual production, you can see the trend line had its first significant dip in 1915 at the 700-metric-tonne-level. That dip bottomed at 500 metric tonnes in 1922. From there, the trend line rises to 600 tonnes in 1925, and flattens at that level.

Then, in the late 1920s, the trend lines again moves upward and takes off in the 1930s, rising to 1,300 metric tonnes in 1940.

For obvious reasons, the World War II years were not favorable for world gold production. Records show annual produced tonnage dropped sharply in 1940 and 1941, from 1,300 metric tonnes in the early war years, down to just under 800 metric tonnes in 1945.

After WWII, annual produced tonnes rose at a slower pace. Tonnage rose in a saw-tooth pattern to just under 1,000 tonnes by 1955. However, by 1970, the annual figure hit almost 1,500 tonnes.

After downturns in 1970-1980, production rose sharply again, moving from 1,200 metric tonnes in 1980 up to more than 2,500 metric tonnes in 2000.

(A good source of trends and other gold numbers is the Gold Institute [www.goldinstitute.org]. The Gold Institute is the international industry association that represents companies that mine and refine gold, and manufacturers of gold products, the bullion banks, and gold dealers.)

Gold Production Around the World

If you examine gold production by areas of the world, you find that, like many things in nature, not all areas are equal.

According to the Gold Institute report on World Gold Mine Production 1999-2003, the top five gold producing countries were South Africa, the United States, Australia, Indonesia, and Canada.

In the period 1976-2000, South Africa led all countries in gold production, with a total of 488.9 million ounces (Moz), according to the Gold Fields Mineral Services Ltd. During that time, gold producers in the United States racked up 155.7 Moz, followed by Russia (the former USSR and former Number 2 in production) with 137.7 Moz, Australia with 122.1 Moz, and Canada with 93.7 Moz.

Gold Fields Mineral Services reported gold producers in the remainder of the world produced a total of 456 million ounces between 1976-2000.

Hands down, South Africa, with some of the deepest gold mines in the world, continues in its long-held position as the world's leading gold producer. Data available for the year 2000 shows world production of gold hit 2,573 metric tonnes, and South Africa led all countries by producing about 440 tonnes (14.1 Moz).

According to the Gold Institute, North America produces almost 20 per cent of the world's gold, and total uses of gold in North America account for 13.5 per cent of world gold consumption (Year 2000 numbers).

Total U.S. gold production, according to the Gold Institute, was

155.7 million troy ounces from 1976-2000, or about 10.7% of the world total of 1,454.35 million ounces for that period. (According to the trade association, the U.S. gold industry is responsible for more than 66,000 jobs, creating more than $6 billion in economic activity and individual wages of nearly $2 billion. Until 1980, the U.S. was a net importer of gold, but today it is a net producer.)

Over a longer period, Goldsheetlinks.com reports, the U.S. production of gold averaged just over 11 million ounces per year since 1997. In 2000, the U.S. produced 12.8% of the world's total of 2,573 metric tonnes. South Africa produced 16.6% of the total.

The top five gold mines in 2000 were in Indonesia, Peru, Uzbekistan, the U.S., and South Africa, according to a 2001 report in the Mining Journal Ltd., *World Gold* (London).

In Indonesia, Freeport-McMoRan C&G led the field, with 2,363,000 troy ounces from its Grasberg Mine. Newmont Mining's Yanacocha Mine in Peru with 1,803,000 ounces was number two. Navol Mining's Muruntau Mine in Uzbekistan, was a close third, with 1,800,000 ounces.

Number four was Barrick Gold's Betze Post Mine in the U.S., with 1.64 million ounces, and the fifth largest producing mine in 2000 was Gold Fields' Driefontein Mine in South Africa, at 1.394 million ounces.

If you divide the top mines according to regions, we see the U.S. and South Africa led. Each country had four of the top 20 producing mines in the world in 2000.

Not only does gold production vary widely by country, it understandably varies within countries. In the United States, for example, most gold production is in the West.

Nevada led all U.S. gold-producing states with 8.4 million ounces in 2000, according to Dr. John L. Dobra's calculations printed on the Gold Institute website (www.goldinstitute.org). States placing a distant second and third in annual gold production at the turn of the century were Utah, with 700,000 ounces, and Alaska, with 546,000 ounces.

South Africa, which is seeing a long-term decline in production in the last 20 years, still manages to produce from one-fourth to one-third of the world's annual gold production. In a paper on gold production reporting up to 1994, John M. Lucas (http://minerals.usgs.gov/minerals/pubs/commodity/gold/300494.pdf) says gold in South Africa has ranged from a high of 1000 tonnes (32 million ounces) in 1970 to a low of 601 tons (19.2 million ounces) in 1991.

Worldwide production trends change slowly, but some new players are moving into the "top producing" statistics. Gold production also fluctuates by area because of various political and economic factors.

According to the Goldsheet website (www.goldsheetlinks.com/production.htm), production in 2000 in the United States, Australia, and Canada dipped slightly, and China's production increased in 2000. Listed as "other significant" producers in 2000 were Chile, Indonesia, Peru, Russia, and Uzbekistan. (Mexico, an emerging gold producer, was recently among the top 20 producing countries.)

Gold Mining Companies: The Majors and the Juniors

Gold stock analysts divide gold producers into "majors" and "juniors," and some refer to some developing juniors as "intermediates."

Majors are the established gold companies with higher capitalizations and revenues, which churn out the most gold tonnage. They hedge and de-hedge (see below) when market conditions dictate, and survive the peaks and valleys in the gold market because they have more access to funding through the capital markets (debt and equity) and hedging. As a result their stock prices are more stable than smaller and un-hedged precious metal companies.

A hedge is a financial asset that offsets losses. Hedges are not used to make a net gain or profit. Instead they are used to protect existing assets.

If a company needs operating funds, and projects a consistent production of a set amount of gold, it can sell the future production to investors/speculators at a set price for each ounce.

If the gold price moves down, the speculator might lose, or hold onto the gold for future appreciation. If the price goes up, the speculator gains as the company that hedged must deliver the quantity promised at its agreed price. While the company might miss its guess on the market, hedging allows the companies to continue with gold production. Hedgers can stop hedging when prices in the open market rise and they can sell gold for better prices.

In recent years, the top three gold producing companies in the world have been Newmont Mining Corporation, AngloGold LMT, and Barrick Gold Corporation.

In early 2002, Newmont Mining Corporation displaced AngloGold and became the world's largest gold producer when it acquired control of Normandy Mining Limited of Australia. Before that acquisition, Newmont had beefed up its holdings by adding Franco-Nevada Mining Corporation Limited of Canada. AngloGold produces about 6 Moz a year and has mining interests in several countries, including the U.S., South

Africa, Australia, Mali, Tanzania, Namibia, Argentina and Brazil.

The merger of South Africa's Barrick Gold Corporation and Homestake recently created one of the world's leading gold producers, with development and mining operations in the U.S., Argentina, Australia, Canada, Chile, Peru, and Tanzania.

Juniors normally have fewer producing mines and aggressively explore and develop gold claims, hoping to strike it rich and survive in a tough market. Juniors and intermediate companies rely more on internal financing, initial public offerings, and private placements to launch themselves or develop promising gold fields.

Because many have less financial ability to survive the ups and downs in the gold market, some juniors survive by selling their gold holdings (leases and mines) or portions of their future gold production to the majors.

One example of an aggressive junior mining company is Wheaton River Minerals (WRM Toronto). In 2000, it bought 100% interest in the Red Mountain Project, an underground gold mine in British Columbia, Canada. A year later it sold it to National Metals Corporation. They also optioned another property to Kinross Gold Corporation, among other deals.

Since then, Wheaton completed a $120 million (Canadian) equity financing, acquired Los Filos Gold Deposit in Mexico and announced plans to acquire the Amapari Gold Project in Brazil.

In addition, Wheaton's three-month earnings ending September 2003 jumped to almost $15 million from $1 million in the comparable period in 2002. The company announced in late 2003 that it expects its gold production to hit 900,000 ounces in 2006.

It's that kind of rapid growth that makes the juniors exciting to speculators.

Into the Future

As with most things, economic demand—the desire for a product and the ability to pay—drives production of minerals, as it does many agricultural products and other commodities.

Historically, demand for gold came from people who wanted it as adornments. Later it was used as a medium of exchange. In more modern times, gold has been used by countries to support paper money and banking operations, which found gold overly cumbersome in promoting commerce. (And today, there are questions about paper—with electronic banking and blips on computer screens moving to replace paper.)

Other than the mask of King Tut—the Boy King of Egypt—perhaps one of the most recognized works in gold is the 32 cm high (little more than 12 inches high) World Cup, which is 18-carat gold. (See the World Gold Council and its Amazing Facts at www.gold.org and www.gold.org/discover/knowledge/amazingfacts/index.html.) One of the most unusual uses of gold in recent times was to plate the doors of a bank in the small country of Oman in the Arabian Gulf.

In overall demand, gold is low compared with other metals. Industry uses much more platinum, palladium, and silver. Gold's use in electronics and dental work account for about 12 percent of total gold demand.

But gold is also high on the list of recyclables, and was so even before recycling became a fad and then a routine. In fact, your gold crowns or perhaps your wedding ring may be made partly of gold that was first mined in prehistoric times. The World Gold Council says: "Today, at least 15% of the annual gold consumption is recycled each year."

Besides mundane recycling that comes from reclaimed dental work, old jewelry and other things that are recycled, there is also more "exotic" recycling:

- Fortune hunters find bags of gold coins and gold bars when they enter a small cave in a remote part of South or Central America. Newspaper articles and a book follow.
- Archaeologists find a pharaoh's horde of gold in an Egyptian tomb. The discovery provides the basis of a movie.
- Scavengers using modern equipment find gold bars, old coins, and plates of gold while salvaging in the deep seas. It is a major find, the remains of an unlucky ship that did not survive the trip back to Europe from the New World after Columbus. The ship was one of many that reportedly went down during the period. Adam Starchild, in *Portable Wealth: The Complete Guide to Precious Metals Investing* (Paladin Press; January 1998, p. 4) calls it "the greatest surge in world gold production." The discovery gets worldwide attention.

These are all examples of "recycling" of gold—old gold that is brought once again into circulation.

People may have a better understanding of the uses of gold production years ago than they do today because of its former limited uses—jewelry, money, and dental work. But high tech is changing all that, creating a whole new world of uses and demand for gold.

Here's a little glimpse of how the use of gold has expanded in recent years:

- 1935: Western Electric finds that its Alloy #1 (a mixture of 69% gold, 25% silver, and 6% platinum) is just right for *all* switching contacts for AT&T telecommunications equipment. (Source: www.goldinstitute.org/history/)
- 1947: Gold contacts go into the first transistor, the building block of electronics.
- 1960: Lasers use gold-coated mirrors to maximize infrared reflection.
- 1969: Rocket scientists coat visors with gold to protect U.S. astronaut's eyes from bright sunlight in the Apollo 11 moon landing.
- 1986: The first gold-coated compact discs.
- 2000: Astronomers in Hawaii used large gold-coated mirrors of the Keck Observatory's twin telescopes to give the most detailed ever images of Neptune and Uranus (see www.goldinstitute.org/history/).

Today, gold is used in an ever wider range of important commercial and industrial applications. In addition to electronics, aviation and cars, gold is now being used in telecommunications, lasers and optics, and medicine.

Why?

The answers lie in gold's many attractive attributes. They are just right for some of our most advanced technologies. These attributes include:

- ductility and malleability (it can be shaped into almost anything!)
- resistance to corrosion and biocompatibility
- electrical and thermal conductivity
- high reflectivity

You now find gold in smart-cards (gold plating and bonding wires), automotive electronics (ignitions control, electronic fuel injection, anti-lock brakes, and air bags), sensors, medical implants, and drug delivery systems.

You'll also find gold in communications technology (battery connections for mobile phones), gold-plated edge connectors in computers, and gold bonding in semiconductors.

Gold is also being used in nano-technology—in nano-wires, nano-electronics, and sensors.

In medicine, because of its qualities of permanence, conductivity, general compatibility with the human body, and its opaqueness to X-rays, gold is in greater use in medical devices. These include wires for pacemakers, gold-plated stents used to inflate and support arteries in treating heart disease, and implants in areas of increased risk of infections.

And gold is being used in drug-delivery microchips that can deliver drug dosages on command. Gold is even being studied for possible use in treating some forms of cancer.

Next up, expect to read more about gold as a chemical catalyst, including use of gold in the areas of pollution control and fuel cell applications.

But gold is being used in more than just technology. It's also being used internally in colloidal form, as it was in the early days of China, where gold was used in the treatment of such things as small pox and measles.

Pure gold suspended in pure water makes a colloid (like colloidal silver). And colloidal gold has been used for centuries as a medication. Accounts dating back to the Middle Ages tell of colloidal gold's use in treating anxiety and impatience, controlling pain, and general health and restorative properties.

Europeans have used "over-the-counter" colloidal gold and "gold-coated pills" for "healing activity of the heart" and improved blood circulation for more than 100 years, according to an article on www.colloidalgold.com/past.htm.

In the U.S., colloidal gold and gold salts have been used since the 1920s for treatment of such ailments as rheumatoid arthritis. Recent research describes beneficial effect of colloidal gold on cognitive functions and also on the tenderness and swelling of joints of rheumatoid arthritis patients.

Corti and Holliday (*Materials World*, February 2003, p. 14) add:

> "From the treatment of serious diseases and ailments, the ability to communicate faster, and the protection of key components in space, to the future provision of pollution control systems, chemical processing plants and clean energy generation, gold has a growing importance in people's everyday lives. Having been revered throughout history, the requirement of new manufacturing techniques, changing consumer demands and emerging new products are set to put gold center stage once again."

All these uses—and future developments—underpin and help drive up the production of gold. (For an impressive timeline of gold's use, check out the Gold Institute's web site, specifically www.goldinstitute.org/history/).

So today, gold has proven to be much more valuable and versatile than simply as a medium of exchange, a storehouse of value, and a basis for jewelry.

The World Gold Council, which tracks trends in demand for gold, divides demand for gold into four categories: jewelry, net retail investment, industrial, and dental demand.

Total demand in 1996 was 4,189.6 tonnes. Jewelry's demand of 3,311.0 tonnes was strongest. Retail investment gold demand was a distant second at 459.4 tonnes, followed by industrial demand at 349.4 tons, and dental demand at 70.1 tonnes. Gold Field Minerals Services (GFMS) lumps three groups into net retail investment. These are:

- short-term speculators, such as those buying and selling on the COMEX
- gold producers that hedge (sell gold forward at a fixed prices) and then de-hedge by selling gold as the price of gold moves upward
- what the industry calls "bar hoarders"—those who hold bullion for longer-term appreciation.

Since 1996, demand has trended down somewhat. According to the World Gold Council numbers, demand slipped across the board to 3,800.3 tonnes in year 2000 and sank to 3,414.5 tonnes in 2002. (See World Gold Council Gold Trends, www.gold.org/value/markets/Gdt/index.php and Gomes's website at www.gfms.co.uk). Some of that downtrend probably had to do with the price of gold, which had been in a slump.

The statistics in this chapter also lead us to an interesting point. Gold production has risen back up to around 2,500 metric tonnes a year, as of year 2000. However, even though demand slipped somewhat, it is now holding around 3,500 tonnes a year. How will this work out?

Supply And Demand

To see this situation in detail, we can look at the "investment case for gold" by GoldEx (www.goldex.net/whygold/gold-fundamentals.html). GoldEx says the annual demand for gold (which it sets at about 3,500 tons a year), has been "increasingly and significantly higher than mining production (about 2,500 tons a year)," and the gap between production and demand grows wider.

According to GoldEx there is a gap between supply and demand, which has consistently widened since 1988—from 150 tons in the late 1980s to more than 1,000 tons in 1996.

Now, given that the shortfall continues to widen, how does the demand side keep up each year? We find the answer in three elastic factors that come into play in the market: "official-sector sales," "old gold scrap," and "implied net disinvestment."

Consider their impact on figures for 2001 and 2002. Gold production from mining operations slipped a bit in those two years, down to 2,623 tonnes in 2001 and to 2,590 tonnes in 2002, according to Gold Fields Mineral Services Ltd. and the World Gold Council.

However, the total "supply" of gold rose to 3,900 tonnes in 2001 to meet demand because the world's central banks and other institutions that hold gold in "reserves" sold 529 tonnes into the market, and sellers of "old gold scrap" provided 706 tonnes to the supply side. The other 42 tons that make up the supply total came through disinvestment—people selling in the markets.

Year 2002 was pretty much a repeat of the supply/demand pattern of 2001. The total "supply" of gold rose to 3,984 tonnes because the world's central banks and other institutions sold 559 tonnes, and sellers of "old gold scrap" provided 835 tonnes. (More people invested in gold in 2002 than the previous year.)

Some gold experts and mining company leaders have expressed concern about the persistent failure of mining companies to produce enough gold to meet demand.

Twenty years ago, some writers and mining leaders, including Bill Brown, CEO of Goldfields, wrote in the *Engineering Mining Journal* that gold production was no longer a viable business (Doran: 2000). But times change and uses and demand of gold move on.

In a speech in early 2003, Barrick vice president for exploration Alex Davidson warned that the gold mining industry had invested so little in exploration that the current reserves could be depleted in 10 years.

"Given that most mining projects require 5-8 years from discovery to production," he said, the industry is not currently funding exploration at a level to replace reserves.

He says the juniors, where the action will be in coming years, are a better investor vehicle than the majors, which will be scrambling for gold fields to produce.

Which leads us to an interesting question: how much gold is left to be recovered?

The answer is in the next chapter!

CHAPTER 7

How Much Gold Is Left to Be Found?

Determining the amount of gold yet to be discovered on our little planet is a monumental task, to say the least. And there are all sorts of things to be considered that will affect how well we can determine that number, and how meaningful that number is—from new mining techniques to advances in exploration to the potential profitability of each new discovery.

It really is a story in and of itself. And so we'll begin at the beginning.

Gold by Grade

To date, humankind's mining efforts have managed to take 142,600 tons of gold from the ground. That's in the last 6000 years. For a visual, you would be able to fit all that gold in your favorite basketball arena without stacking it more than 30 feet high.

Early in our history we mined gold in its high-grade form because that is what we were able to do. What qualifies as high grade depends on the type of deposit and its location. It also depends on the company that has made the discovery. In one situation the cut-off for gold to be high grade might be 6 grams per metric ton, as in Homer Gold's discoveries in Canada; in another, 1 gram per metric tonne might be used as a cut-off as in Chariot Resource Limited's recent discoveries in Peru. Generally gold deposits above 4 grams per tonne are considered to be high grade.

Because of gold's amazing durability, it withstands natural erosion processes better than the surrounding materials. In other words, while dirt and other rock deposits wash away in the rain and periodic floods over time, gold doesn't, because of its weight and stability. Many early gold miners found deposits of significant size by just scanning the top soil.

Gold also has a tendency to partner up with quartz. Certain types of quartz veins often indicate large high-grade gold deposits. These weren't as easy to locate, but they weren't too difficult either. And in both cases, the mining process was fairly straightforward. Mining gold from quartz veins is a more complicated process, but it didn't require any revolutions in technology.

Now, most typical high-grade deposits indicate there is also a large deposit of low-grade gold in the vicinity, low grade being anything that doesn't make the cut as high grade, but for simplicity's sake less than 4 gram per tonne and often less than 1 gram per tonne. (It really is impressive when you think about it…that mining operations can make a profit extracting less than a gram of gold for every tonne of rock.)

Low-grade gold presents much more of a challenge when it comes to mining. But it's also very enticing because the deposits are often much larger than the original high-grade ore. For example, the Round Mountain mine in Nevada yielded 350,000 ounces of gold from its high-grade gold deposits between 1906 and 1969. When the technology became available, the low grade ore was mined. In the last 25 years more than 4 million ounces have come from the Round Mountain low grade deposits and the miners aren't done yet. It's expected that the Round Mountain low-grade deposit will yield a total of 11 million ounces of gold before it is exhausted.

Until recent history, locating low-grade deposits ranked low as a priority. They were expensive to locate and often proved too costly to mine. But recent advances in both exploration and extraction have made low-grade deposits attractive for their new potential for profit.

Aside from high-grade and low-grade deposits, gold mining is further complicated by the type of gold. Microfine gold, also called invisible gold, may form sizable deposits but the grains of gold are sized in the microns—very tiny. While low-grade gold is mineable, current processes and technology don't offer a way to economically extract microfine gold. But new refining processes may soon change that.

What Does It Cost?

Exploring and mining for gold can easily rack up expenditures in the millions. Today's mines are often low-grade gold deposits and there is a whole series of expenses to be met in order to pull low-grade gold from the ground.

First you have to find it. Then you have to do tests to determine if it will be worthwhile to mine. The typical mining process for low grade gold is an open pit mine, which involves blasting and moving tons upon tons of rock. Open pit mines are often a half mile in diameter—a huge operation.

Deposits well below the surface of the earth that require shaft mines are just as involved and even more expensive.

It is, as you can tell, a complicated process. But for simplicity it can be broken down into three parts—exploration, mining, and processing. (There is also a reclamation process to minimize environmental impact after a mining operation reaches its end.)

Each step contributes its share to the cost of the mining operation. The cost of each process is difficult to break down into an average because the costs of the different phases of mining vary so much depending on the deposit. But according to the World Gold Council, the overall average cost of a mining operation per ounce of gold is $238—a figure that varies widely when talking about a specific mine, depending on the type and location of the mine as well as the grade of gold. The market price of gold and the difficulty of extracting a particular deposit determine how much a mine can afford to produce.

But it's even more complicated than that. Not only does the market price affect how profitable a mine will be and whether certain deposits are worthwhile to mine at all, but mining technology is constantly advancing and sometimes the changes are rapid.

New technologies in exploration, mining, and processing all can have a substantial impact on the overall cost of a mining operation. When new technologies develop, there is an initial cost to implement them, but once they catch on, they change the whole formula of profitability—and that affects the cost of production of known deposits as well as how aggressively new deposits are sought.

What's New in Exploration?

Exploration is the most important component in beginning to determine how much gold there is that we don't know about yet. It is a rapidly advancing field, but a lot of advances have yet to take hold or to fully realize their promise.

The traditional process for finding a low-grade gold deposit begins with core samples. Areas near known sources of exhausted high-grade deposits are preferred because there is often more gold to be found. But geologists also look at other factors, like the geologic history of an area, to decide where core samples should be gathered from.

So, mining companies send someone out to collect a series of soil samples from a likely area. The samples are then analyzed in a lab and if they look promising, exploratory drilling is done to determine the size of the deposit. The whole game can end here. If the deposit doesn't show promise, then the company will look elsewhere. But if there is enough gold to economically begin mining, then they get to work.

With today's rapidly expanding technology it is becoming easier and less costly to predict where gold deposits will be and to some degree, how big they will be. There is a host of new tools the geologist can use to increase a company's odds of finding a worthwhile deposit. In England the Parys Mountain Group, a group dedicated to the development of Bronze Age copper mines at Parys Mountain, provides a thorough explanation of new techniques for exploration:

Lithogeochemistry: Significant mineral deposits including gold are often found in areas of ancient volcanic activity. Being able to determine the volcanic history of an area is a valuable tool and is the aim of lithogeochemistry. Lithogeochemistry uses a complex process (X-ray fluorescence and inductively-coupled-plasma mass-spectrometry) to determine an area's volcanic history, including the geologic environment during the volcanic activity.

Infrared Mineral Analysis: When rocks are altered in a volcanic or similar geologic event, they often create what geologists refer to as a hydrothermal alteration, meaning that those rocks are different from what they once were because of the volcanic or similar event, and therefore they have unique properties. These properties can help identify the type and size of mineral deposits using Infrared Mineral Analysis. With the recent invention of a Portable Infrared Mineral Analyzer, this process has become much easier and less costly.

3D Modeling: Advances in computer technology show promise when it comes to predicting mineral deposits. Accumulated data can be entered in a computer program that analyzes the data and extrapolates a likely 3D picture of the potential size and shape of mineral deposits. These computer programs are still in the early stages when it comes to applying them to mining operations. They require a staggering amount of data, which makes them quite time-consuming to use.

Paleontology: The group at Parys Mountain has seen a good deal of promise in the analysis of paleontological data to provide clues for the overall structure of a site.

Epithermal Modeling: The U.S. Geological Survey, or USGS, uses epithermal modeling, a technique that has proved very useful and accurate. Different types of epithermal gold deposits have characteristic traits as a result of their geologic history. These traits are used to form models that make it straightforward in some cases to predict the depth, grade, and likely location of a gold deposit.

What's New in Mining?

Mining processes have really changed very little. Most mines today are either shaft mines that delve deep into the earth to retrieve gold deposits, or open pit mines that blast out a large crater of rock to extract the gold from it. Open pit mines became common in the late seventies when heap leaching proved to be a cost-effective means of recovering some low-grade gold deposits.

Where technology has helped in current mining operations is in the level of efficiency and safety that these mines operate at. Ongoing refinements to mining technology allow mining companies to dig deeper for remaining high-grade deposits and allow them to manage larger open pit mines.

What's New in Processing?

New developments in processing techniques are expected to make lower and lower grade deposits economical to mine. Currently huge deposits exist that we know about, that are often left over from a previous operation, and that we have accurately estimated, but are nonetheless out of reach because they just can't be mined without losing money. New processing developments may make those potential mines a profitable endeavor. One new technique is particularly exciting: the Haber Gold Process (HGP).

HGP is a proprietary, non-toxic alternative to using cyanide or other harmful chemicals to dissolve gold. The potential advantages of HGP are threefold and very significant. First, government regulations concerning normal processing of gold—particularly the use of cyanide—are growing more stringent and more costly to follow. Second, initial studies show that HGP is a more efficient process than cyanide. It acts more quickly and recovers more gold. One study conducted by PA Technology resulted in a 94% recovery of low-grade gold in just 6 hours using HGP compared with a 91% recovery in 24 hours using cyanide processing. Finally, HGP has shown great potential when it comes to microfine gold deposits. In most circumstances, cyanide processing cannot efficiently recover microfine gold. Early studies suggest HGP can. Needless to say, HGP will have a substantial effect on the gold industry.

A Partial Survey

So, how is all this relevant to our original questions? How much gold is yet undiscovered? And how much gold do we know about that new processes will make accessible in the near future?

Advances in exploration continue to change how quickly we find new gold deposits and how accurately we can predict the size and location of those deposits. So it is useful to see how much is currently going on in that field. The same is true of processing when you look at known gold deposits that are not yet economical but may become so.

The question of how much known gold will become economic to mine that wouldn't yield a profit in today's environment is the simpler question to answer. Various sources, from the USGS to the World Gold Council, agree that there are 100,000 tons of unmined gold that we know about, meaning we know its location and its grade, we have a good idea of its tonnage, and we can make a useful analysis of what would be required to extract it. 100,000 tons is an estimate, but it's a widely agreed upon figure.

The U.S. Geological Survey further estimates that there are 42,500 tons of gold that can be considered reserves. In other words, 42,500 tons can be mined profitably at today's price and with current technologies. We know of an additional 57,500 tons of the gold that require new technologies or a higher gold price to be considered economical to mine. (Source: 2003 Gold Commodity Summary prepared by Earle B. Amey of the USGS)

The question of how much gold remains undiscovered is really an enormous one. And the truth is, no one knows for sure. There isn't an organization even ready to venture a guess.

You see, determining that figure requires a lot of manpower, a lot of resources, and quite a bit of time. There aren't a lot of companies or organizations with enough capital to come up with an estimate.

There is one entity, however, that does have the means and the motivation to do a systematic and global analysis of precious and industrial metal deposits, including gold.

But first, the work that goes into coming up with such an estimate is significant enough that it merits some recognition.

Epithermal models provide the best means to make a systematic analysis of likely but unknown mineral deposits. The process is begun by examining landmasses and dividing them into regions with a similar geological history and that show broad characteristics of mineral deposit models. These regions are then subdivided into smaller tracts that are more closely grouped based on epithermal models. Further analysis is done of the smaller areas to predict the likely size and type of gold (or other mineral) deposit.

There are five models commonly used to predict gold deposits. They are named Hot Spring, Creede Veins, Comstock Veins, Sado Veins, and High Sulfidation. Hot Spring type deposits typically produce a large tonnage of low grade gold near the surface—ideal for an open pit operation. Comstock, Sado, and High Sulfidation deposits are typically lower in tonnage but higher in grade with High Sulfidation type deposits being the least costly to mine of the three. Comstock type deposits are typically rich in silver as well, as are Creede type deposits (so much so, that they are considered silver deposits). The Creede veins usually hold a low tonnage of relatively low-grade gold.

Now you may have already guessed who would take on the task of predicting undiscovered mineral resources on a global scale, and who would have the resources. It is of course the United States government—the U.S. Geological Survey working with other organizations and universities from around the world.

In 1996, the USGS completed a similar project, but it was limited to resources available in the United States. It was the first such undertaking of its kind, and has provided much useful information. Enough, in fact, that they began to determine the feasibility and practicality of doing a similar analysis of undiscovered global resources. This project is moving forward, but it is in its early stages. The survey plans begin with an assessment of undiscovered copper, platinum, and potash. It will be many years before we have a firm, accurate estimate regarding gold.

What the U.S. Geological Survey did discover concerning gold in the United States is that there are an estimated 14,000 tons of undiscovered gold in this country at a 1 kilometer depth or less. For comparisons sake, our known but unmined resources total 36,000 tons—that includes both deposits that are economic to mine and those that aren't.

Where does that leave us for a global estimate?

If we assume that the nature of the U.S. markets allows for a more thorough search and estimation of known and unknown gold deposits, then we can extrapolate that total undiscovered gold deposits outside of the United States should be at least 23,700 tons within a 1 kilometer depth. (This is done by comparing estimates of undiscovered gold to known resources within the United States and extrapolating that a similar relationship may exist between the undiscovered gold and known resources in the world).

So in total, we would be surprised to find less than 37,700 tons of gold that hasn't yet been discovered. It very likely will be a larger figure. Combined with known resources, we can determine that the amount of gold left undiscovered and/or unmined is nearly the same as what has been mined in the last 6000 years.

There are two factors that make that figure significant. One is that the easiest gold has all been found. As we work our way through the remaining gold, it will become more and more difficult to find and recover.

Second, the rate of gold production has increased dramatically over the years. We said in Chapter Six that 50 percent of all the gold ever produced in the world has been produced since 1960—in just over 40 years.

Given that we now produce gold at the rate of 2,500 tons a year, five times as fast as in 1900, that remaining 50% of the world's gold supply will go pretty fast. So simply from supply and demand, gold will become more and more valuable as we move toward the exhaustion of the world's gold supplies.

And finally, the demand for gold is growing constantly.

All these factors contribute to our position that gold is a crucial part of any investor's safety net. And it will become more and more important as time goes by.

Part III

Modern Economics and Gold: The Pressure Mounts

CHAPTER 8

Central Banks and Gold

In a panic, the trader stared at his computer screen. The central banks had unexpectedly announced a multi-year schedule of gold sales, in an unprecedented inter-bank agreement. Now the market was melting down, catastrophically, and he couldn't get out of his position. His broker's website said, "Due to the overwhelming volume of trades in the gold market, limit orders will not be accepted. All orders must be placed at the market, and immediate fulfillment of your order may not be possible. Thank you for your cooperation..."

He had called his broker and screamed, Sell it all, now! The broker had sounded dazed, and eventually just asked: To whom?

The time was late September, 1999. An agreement on gold sales had been announced: the signatories were the European Central Bank and the central banks of Austria, Belgium, Finland, France, Germany, Ireland, Italy, Luxembourg, the Netherlands, Portugal, Spain, Sweden, Switzerland, and England. The United States, Japan, and Australia, although not signatories, had said they would also follow suit.

Similar announcements by central banks had been made before, although not by as many countries together (all told, about 85% of the world's official gold holdings were covered by the agreement, whether formally or informally). This agreement was unprecedented.

Central banks can exert a tremendous influence on global gold markets. In this one announcement, the market was thrown into total chaos. But not the kind you're probably thinking...

In this agreement, the banks had agreed to a **moratorium** on gold sales for the next five years, beyond what had been previously scheduled.

The gold market surged the next day, going from $268/oz. to $284. When the markets opened again, it went to $299...then exploded up another 9% to $328 in four and a half hours. Short sellers were crushed instantly; even more unfortunate were those who had sold calls, or bought puts (like the trader described above), as the leverage worked against them. Some gold call options went from $200 to $10,000 in a matter of hours (yes, going up by a multiple of 50!) Gold advocates who were long gold made a fortune overnight: traders who had bet against gold were wiped out in an equal amount of time.

But what are central banks? Why do they have such tremendous influence in the gold markets? Why do they have so much gold (32,000 tons or so), and more importantly—what are the chances they'll decide to sell this gold and crater the market?

Central banks have been around since the formation of the short-lived Swedish Riksbank in 1656. Today, they include such prestigious institutions as the Bank of England, chartered in 1694. During the gold standard era, central banks had a different role than today. They were unable to inflate their currencies willy-nilly as they can now, so their role was more limited. Today, their role has expanded tremendously, and they have a dramatically greater influence on national economies.

To understand the modern role of central banks, and their reasons for holding gold, we have to review a little history. It all started when the Japanese attacked Pearl Harbor in 1941…

* * * * * * *

> *"In 1918, most people's only idea was to get back to pre-1914. No one today feels like that about 1939."*
>
> *British economist John Maynard Keynes, during the planning of the post-World War II global economic system*[33]

When the United States was pulled into World War II, her leaders immediately began planning for the international post-war economy.

At the time, many claimed that the recent problems in the world could be attributed to the gold-based economic system then in place. They blamed the "golden straitjacket" for the Depression, the runaway inflations in various countries, the hostile trade policies between nations, and even the war itself.

Of course, this is nonsense. As we saw in Chapter Four, the gold standard promoted international stability. Most of the problems attributed to golden economics were in reality the result of World War I, from the devastating wartime overspending on one hand, to the vindictive Treaty of Versailles on the other. Nevertheless, the politicians found it easier to blame gold than themselves.

On December 14, 1941, exactly one week after Pearl Harbor, the Secretary of the U.S. Treasury (Henry Morgenthau Jr.) directed his assistant to prepare a memo on "postwar international monetary arrangements." That assistant, Harry Dexter White, eventually created the plan for the organization known today as the IMF (International Monetary Fund).

White did not work alone. He began corresponding with the prominent British economist John Maynard Keynes, who was then an advisor to the British Chancellor of the Exchequer. Together the two men drafted a plan for an international organization that would fix currency exchange rates between member countries, and remove from those currencies their direct links to gold.

Gold's removal from the monetary system was quite deliberate—Keynes was very explicit in his disdain for the yellow metal. He said, "I have spent my strength to persuade my countrymen and the world at large to change their traditional doctrines...Was it not I, when many of today's iconoclasts were still worshipping the Calf, who wrote that 'Gold is a barbarous relic'?"[34]

The IMF was to serve as a sort of global credit union. Member countries would pay a fee (a "quota") upon joining, and 25% of the quota had to be paid in gold. National currencies would no longer be linked to the yellow metal, except the U.S. dollar which retained its convertibility ($35 was equal to one ounce of gold). All nations (other than the U.S.) would set exchange rates for their own currencies in terms of the dollar, and their central banks were obliged to manage their economies so as to maintain these rates (although they could change the rates occasionally if need be, with the Fund's approval).

A member nation which ran into economic trouble could withdraw its 25% gold quota from the Fund for emergency use, to be repaid later. If more was needed, a member could also borrow up to 3 times its quota, as long as it adhered to a Fund-approved economic reform plan.

In July 1944, 730 delegates from 45 countries met in Bretton Woods, New Hampshire, three weeks after the invasion of Normandy. They drafted and signed the Articles of Agreement for the Fund—and the "Bretton Woods system" was born.

The new system made the dollar the centerpiece of global finances. This made sense, as the war was ruining the economies of most of the participants, with entire cities being bombed into rubble—whereas the U.S. economy was only growing stronger (total production in 1947 was 30% greater than that of 1941). In addition, by the end of the war the U.S. had about 75% of the world's supply of monetary gold, so the dollar had strong backing.

Bretton Woods relied on the dollar as the anchor for all other currencies. In turn, the dollar was anchored by gold. If the latter bond held, then the former could be maintained as well. Conversely, if the dollar's tether to gold was broken, then the system would collapse. Thus, in a sense the entire global system relied on the discipline of U.S. leaders to restrain themselves, and not create excessive dollars relative to the gold that backed them.

Of course, it's never wise to rely on the fiscal discipline of politicians. The U.S. found itself in a position very much like the unregulated banks of earlier history: it had a lot of gold in its vaults, and "deposit slips" or "receipts" for that gold (i.e., dollars) were in wide circulation. However, the "depositors" (dollar holders) rarely exchanged their receipts for gold, as the receipts were widely accepted in trade and were much more convenient to use than the heavy, bulky metal itself. Of course, when the receipts are broadly accepted as tender and can be spent as money, the "banker" then becomes sorely tempted to print up some more for himself...unbacked by actual gold in the vault.

So a few years later, just like those unregulated banks, the U.S. was issuing new receipts...first a trickle, then a torrent. The Marshall Plan in Europe, the occupation and rebuilding of Japan, global militarization against encroaching Communism, massive social spending at home, the Korean War—all required financing. New dollars were created and flooded the world, far above any possible convertibility into gold.

Of course, many in the world noticed this spending spree and started cashing in their receipts. The national gold stock began to shrink—the U.S. lost $3 billion of its gold just from 1958 to 1960.

By 1960, the dollar assets in foreign hands alone would have exhausted the U.S.'s entire stock of gold.[35]

* * * * * * *

> *"We have gold because we cannot trust Governments."*
> *Herbert Hoover*

> *"The more that governments tried to find wiggle room around the constraints of the Bretton Woods system, the more the public and the speculators followed Hoover's dictum and turned to gold as the ultimate hedge against the irresponsibility of governments."*
> *Peter Bernstein, economist and historian*[36]

By the early 1960's, the failure of the U.S.'s fiscal discipline was widely recognized. No longer was it only discussed by government officials behind closed doors—in 1960 the markets themselves were pricing gold as high as $40 an ounce, 14% above the official price. This was a clear recognition that the dollar was depreciating against gold, and an implicit accusation that the international agreements were not being observed.

Political leaders had two choices. They could reduce the number of dollars circulating in the world, thus adhering to their obligations to maintain dollar-to-gold convertibility. However, this would also require a corresponding cut in spending, and the U.S. was just entering a decade of expanded military involvement in Vietnam, along with a massive increase in social programs.

The second option was to suppress the price of gold through market intervention, thus silencing the market's accusations (although not solving the underlying problem). Of course, this was the option selected.

In November 1961, the central banks of Belgium, France, Italy, the Netherlands, Switzerland, West Germany, the U.K., and the U.S. began to work together to manipulate the open-market price of gold. Known as the "gold pool," these banks would sell gold into the market whenever the price was significantly above the official price of $35/ounce. For the first few years, this tactic worked fairly well.

However, as the 1960's wore on, U.S. spending was increasing on a massive scale. President Johnson's War on Poverty and his Great Society, when combined with dramatically increasing costs in the Vietnam War and Johnson's refusal to raise taxes to pay for it all, required a flood of new dollars. The markets naturally commented on this profligacy by buying gold...and the gold pool had to work feverishly to squash the price.

Gold started pouring out of Fort Knox into the markets. From late November 1967 to early March 1968, the U.S. lost nearly $2.5 billion of its gold in the gold pool.[37] Finally, political leaders had to admit defeat.

As one economist puts it, "On March 17, 1968, after seven fruitless years of pouring their treasure into the maws of the speculators, the members of the gold pool decided the game was up. They announced that henceforth they would no longer supply gold to the London market, or any gold market...From that point forward, the price of gold in the free market would be left to private parties to determine."[38]

A year later, the open-market price of gold was over $43. The market now judged the dollar to be worth 23% less than what the U.S. government claimed.

The United States had managed to squander a national wealth that was unprecedented in world history. After World War II, the U.S. had 75% of the world's gold. By the end of the 1960's, it had less than 30%.[39]

What was worse, the U.S. was still printing billions of additional unbacked "receipts" for which it was obliged to provide gold, upon request. Although this obligation was limited to treasuries and central banks rather than individuals, nevertheless there was now a tremendous temptation to buy gold from the U.S. at $35 and sell it on the open market for $43 for a quick, guaranteed profit. And the more dollars that were printed, the higher the price would go, and the more profit there was to be made. It seemed obvious that the U.S.'s gold wouldn't last long under these circumstances.

Clearly, a crisis was at hand.

****

"The kind of transcending value attributed to the dollar has lost its initial foundation, which was possession by America of the greatest part of the world's gold." [40]

General Charles de Gaulle, President of France, February 4, 1965

By October of 1967, the U.S.'s gold was down to $12 billion while liabilities stood at $33 billion. French economist Jacques Rueff described the U.S.'s inability to pay off its creditors in gold: "It is like telling a bald man to comb his hair. There isn't any there." [41]

The situation worsened each year. By 1971, U.S. liabilities stood at $60 billion, while the stock of gold was down to $10 billion.[42] Of course, the owners of these liabilities were unhappy with this underlying rot in their dollar assets. Charles de Gaulle, the President of France, had already called for a return to the international gold standard. Other French leaders added that the present system was "absurd...a serious obstacle to social progress" and pointed out that a return to gold would "stop the decay of the world money system." [43]

The U.S.'s ultimate response is not really surprising...

The final straw came in early August of 1971, when a British economic representative came to the Treasury in person and asked for $3 billion in gold. A few days later, on August 15, 1971, President Nixon announced that he had "closed the gold window." From that point on, dollars would no longer be converted into gold.

In one short announcement, the United States defaulted on some $60 billion in international obligations. Of course, President Nixon presented it a bit differently. He said, "I have directed the Secretary of the Treasury to defend the dollar against the speculators...Now the other nations are economically strong, and the time has come for them to bear their fair share of the burden of defending freedom around the world. The time has come for exchange rates to be set straight...There is no longer any need for the United States to compete with one hand behind its back...We are not about to ease up and lose the economic leadership of the world."

As Paul Samuelson, the Nobel Prize-winning economist wrote in the *New York Times*: "The President had no real choice. His hand was forced by the massive hemorrhage of dollar reserves of recent weeks...For more than a decade the American dollar has been an overvalued currency."

The Bretton Woods system was now officially over. But what would replace it?

****

"Collapse of the Bretton Woods system in the 1970s came as a shock, but it was like the shock of the death of a family member who had been ill for a long time."

Margaret de Vries, IMF economist and historian[44]

"One American who wanted to buy a loaf of bread in Paris and offered a dollar bill to the baker was told, 'That's not worth anything any more.'"

The New York Times, *August 17, 1971*[45]

Gold was now set free from any official entanglement with the dollar. The markets soon expressed their collective opinion about its true worth.

At the beginning of 1972, the yellow metal stood at $46 per ounce. In 1973, the price had more than doubled to $100. By 1978, it was $244. In 1979, $500.

This didn't happen in a vacuum, of course. Much of it reflected a crisis of confidence in the now-floating dollar. Inflation in the U.S. skyrocketed, hitting 8% in 1978, and 12% in 1979.

Meanwhile, rather than clinging to gold as a lifeline, the U.S. Treasury was distancing itself from it. The Treasury auctioned off some of its gold in 1975, 1978, and 1979, because "Neither gold nor any other commodity provides a suitable base for monetary arrangements."[46]

Also during this period, the IMF was struggling to redefine itself. On April 1, 1978, the 2nd Amendment to the IMF's Articles of Agreement finally removed any formal role of gold in the international monetary system. (Actually, President Nixon had already done that 7 years earlier, but this at least made it official.)

However, this left the IMF with a fundamental problem. What good is an organization designed to administer a world system based on gold… when the gold itself is removed?

While its members mulled over this question, they also decided to start selling off the IMF's gold.

As of August 1975, the IMF held 153 million ounces. The Fund decided to sell 50 million ounces over the next few years. Half of this amount was "restituted" to member countries, i.e. it was sold back to them at a very low price. The other 25 million were sold via auctions held from June 2 1976 to May 7 1980. The average price: $246/oz.

That left the IMF with about 103 million ounces, the amount it still holds today. The Articles of Agreement allow the IMF to sell more of its gold if desired, but only by a vote with an 85% majority. (More on the IMF in the next chapter.)

Interestingly, the combined auctions of the Treasury and the IMF sold gold into a rising market. Despite a substantial amount of metal being sold over 5 years or so, the price of gold went up by hundreds of dollars over that period. When the market wants gold, even large sales can't stem the tide.

* * * * * * *

"The market shows that people don't trust the governments and they don't trust paper money either."

A trader quoted in the New York Times, *January 4, 1980*[47]

"We're in World War Eight, if you believe the market."

Commodities broker James Sinclair, in the New York Times, *January 22, 1980*[48]

By 1980 the gold market was in a blow-off top; the price briefly hit $850/ounce in January. This was the end of a spectacular decade-long run in which the price of the metal had gone up by more than 19 times the original price.

The bull market in gold reflected a serious crisis in the dollar. As mentioned earlier, inflation was running at 12% in 1979. The dollar was also devaluating rapidly against other currencies.

Finally, the U.S. had to take drastic measures. Paul Volcker, the Chairman of the Federal Reserve, announced a new clampdown on the money supply. Interest rates skyrocketed: the U.S. Prime Rate hit 20%.

Volcker's policies were brutal, or so the commentators complained. Unemployment was already high at 7.1% in 1980, but climbed as high as 10.6% in 1982 before finally falling. Inflation also remained high for several years. But slowly, the economy stabilized. As inflation was finally reigned in, the gold price subsided.

The falling inflation and interest rates also had the effect of making paper assets more attractive, which shifted investments out of precious metals and into paper. Soon, the long bull run in equities would begin.

Precious metals slumped and then were fairly quiet for more than a decade. Central banks sporadically announced gold auctions. From 1987 to 1992, about five hundred tons were sold: then sales picked up over the next seven years, averaging about four hundred tons per year.

Then, in the late 1990's, things suddenly got more dramatic. As described in the beginning of this chapter, in late 1999 the world's major central banks agreed to limit their gold sales. Gold's price exploded upwards 22% in just a few days.

Thus began the modern gold market. There had been conflicting signs before this announcement (the "Joint Statement on Gold") had been made. The Bank of England had announced a sale of 415 tons just four months earlier, and the Swiss had made similar noises before that.

Yet at the same time central bankers still spoke approvingly of the yellow metal. "We are convinced that gold will continue to play a role as a currency reserve, especially in times of crisis," the vice chairman of the Swiss National Bank had said in 1997.[49] The Annual Report of the Bank of France for 1997 included this: "Gold remains an element of long-term confidence in the currency...Above all, holding gold is, from the political point of view, a sign of monetary sovereignty [and] an insurance policy against a major breakdown in the international monetary system."[50]

Indeed, the new European currency, the euro, is partially backed with gold. The European Central Bank, opened in 1998 to manage the euro, keeps 15% of its reserves in gold. Also, central banks across the globe have a vested interest in maintaining the value of gold: with about one-quarter of the entire world's supply in their reserves, central banks would be hurt if the price of gold were to go down too far.

Thus, it really isn't surprising that so many of the world's central banks had decided to unite in their restriction of gold sales. The Joint Statement on Gold pointedly said, "In the interest of clarifying their intentions with respect to their gold holdings, the above institutions make the following statement: Gold will remain an important element of global monetary reserves..."

Of course, central banks will always act in the best interests of their countries. So, in these actions by their central banks, we see that the major Western nations are firm in their desire to support gold.

Why is this so? Central banks manage currencies, none of which are pegged to gold anymore. Why are central banks still keeping large amounts of gold as a reserve asset?

Central Banks and Gold

According to the IMF, central banks hold approximately 29,000 tons of gold and monetary institutions hold another 3,500 tons for an approximate total of 32,500 tons of declared reserves.

The international average of central bank gold to reserves is about 11–12% at current market prices. European Union reserves are over 25% and the U.S. holds close to 60%.

The chart below shows the stability of gold reserves over the last 50 years. As we can see from the chart, total reserves have remained relatively the same:

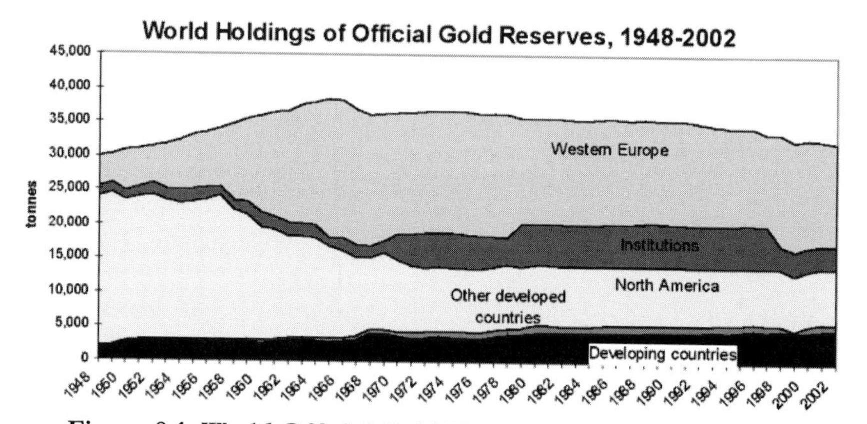

Figure 8.1: World Official Gold Holdings (Sep. '03)
[Chart from World Council Website http://www.gold.org/value/official/reserve_asset/background.html]

The next table lists in detail the 80 largest countries and institutions holding gold. (Taken from the World Gold Council, http://www.gold.org/value/stats/statistics/archive/pdf/sept_2003.pdf. This source contains 116 entries all told.) As we can see, most countries have some gold reserves:

Figure 8.2: The Largest Official Gold Holdings (Sep. '03)

		Tonnes	Gold's share of reserves			Tonnes	Gold's share of reserves
1	U.S.	8,135.4	56.9%	41	Finland	49.0	5.5%
2	Germany	3,439.5	43.4%	42	Bulgaria	39.9	7.9%
3	IMF	3,217.0	N/A	43	Norway	36.8	1.9%
4	France	3,024.8	54.1%	44	WAEMU	36.5	6.4%
5	Italy	2,451.8	45.9%	45	Malaysia	36.4	1.2%
6	Switzerland	1,722.8	31.1%	46	Slovak Rep.	35.1	3.8%
7	Netherlands	842.5	48.6%	47	Peru	34.7	4.0%
8	ECB	766.9	N/A	48	Bolivia	28.3	33.6%
9	Japan	765.2	1.6%	49	Ecuador	26.3	28.4%
10	China	600.3	1.9%	50	Syria	25.9	N/A
11	Spain	523.4	16.7%	51	Morocco	22.0	2.3%
12	Portugal	517.2	45.0%	52	Nigeria	21.4	3.2%
13	Taiwan	422.1	2.8%	53	Ukraine	15.8	2.9%
14	Russia	388.2	6.8%	54	El Salvador	14.6	9.0%
15	India	357.7	4.9%	55	Cyprus	14.5	6.0%
16	Venezuela	346.2	25.4%	56	Korea	14.0	0.1%
17	Austria	317.5	28.5%	57	Brazil	13.8	0.3%
18	U.K.	313.2	8.6%	58	Czech Rep.	13.8	0.6%
19	Lebanon	286.8	23.1%	59	Neths. Ant.	13.1	27.4%
20	Philippines	271.9	19.6%	60	Jordan	12.8	2.9%
21	Belgium	257.8	21.7%	61	Cambodia	12.4	15.9%
22	BIS	193.7	N/A	62	UAE	12.3	1.0%
23	Sweden	185.4	10.2%	63	Colombia	10.2	1.1%
24	Algeria	173.6	6.5%	64	Canada	6.2	0.2%
25	Libya	143.8	7.9%	65	Ghana	8.7	11.3%
26	Saudi Arabia	143.0	7.1%	66	Latvia	7.7	6.5%
27	Singapore	127.5	1.7%	67	Ethiopia	7.6	8.3%
28	South Africa	123.8	18.1%	68	Slovenia	7.6	1.1%
29	Turkey	116.1	4.4%	69	Myanmar	7.2	13.0%
30	Greece	107.2	22.3%	70	CEMAC	7.1	4.5%
31	Romania	105.3	14.1%	71	Guatemala	6.8	3.1%
32	Poland	102.9	3.7%	72	Tunisia	6.8	3.0%
33	Indonesia	96.5	3.2%	73	Macedonia	6.2	8.8%
34	Australia	79.7	3.4%	74	Mexico	5.9	0.1%
35	Kuwait	79.0	9.5%	75	Lithuania	5.8	2.4%
36	Thailand	77.8	2.3%	76	Ireland	5.5	1.7%
37	Egypt	75.6	6.1%	77	Nepal	4.8	4.6%
38	Denmark	66.6	2.1%	78	Bahrain	4.7	2.8%
39	Pakistan	65.1	6.9%	79	Mongolia	3.7	12.7%
40	Kazakhstan	52.8	14.1%	80	Bangladesh	3.5	1.6%

Why Do Central Banks Hold Gold Today?

Since the end of the gold standard, currencies "float" against each other, with exchange rates changing constantly. Each country tries to manipulate its currency's value against the others, for trade advantages. Usually, nations want cheaper currencies, to spur their exports (since it makes their goods cheaper for other countries to buy). So over the long term, there's a consistent devaluation of all currencies, competing with each other in a never-ending race to the bottom.

Politicians don't want the discipline of a gold standard. It would be hard to imagine all the independent economies around the world agreeing to and conforming to a gold standard. They would have to give up their deficit spending, competitive devaluations, and control of their currencies—an unlikely prospect.

Central banks are supposed to keep their national economies stable. This is a difficult task when the national currency is constantly eroding in value. The answer is for the central bank to maintain a "reserve" of something valuable—to beef up the nation's balance sheet, so to speak.

But what can be held in reserve? Each bank can't keep its own currency, which is constantly devaluing...nor will other nations' currencies be any better. The dollar has historically been the reserve of choice, since for most of the 1900s it represented (at least in theory) a certain amount of gold. Now that this relationship has been severed, and the U.S. currency is under severe pressure (see Chapter Ten), the dollar isn't so attractive anymore.

Central banks need to keep a reserve of something valuable...that keeps its value over time...and that is independent of the whims of politicians in its own nation, or in others.

That something is gold.

Central banks have learned that gold has many advantages, making it an ideal asset for their reserves:

1. Diversification—To minimize risk in any portfolio, including central bank reserve assets, it makes a tremendous amount of sense to diversify. Studies indicate that returns of gold have a low or even negative correlation to those of stock and bond markets. When the stock and bond markets are doing well, the gold markets tend to have lackluster returns. But when the bond and stock markets do poorly, gold returns historically outperform. This low- to-negative correlation makes gold an ideal asset with which to diversify.

2. Gold is no one's liability—It is not directly affected by the policy or action of any individual government. It cannot be repudiated or frozen, as can foreign securities.
3. Confidence—Research has shown that in most countries the public takes confidence from knowing that their government holds gold.
4. Income—Historically, gold was thought of as a non-income producing asset. But central banks have now learned to lend out a portion of their gold, which allows for reasonable returns.
5. Unexpected needs—Gold is seen as an "asset of last resort" and fulfills a "war chest" role. In emergencies, countries may need liquid resources. Gold is liquid and universally accepted as a means of payment. It can also serve as collateral for borrowing.

Most gold experts and economists believe the overhang of gold from central banks—i.e., the likelihood of major gold sales by central banks—has been diminished.

> To quote Alan Greenspan in a speech to Congress in 2001, gold is an important asset to central bank reserves:
>
> "[G]old still represents the ultimate form of payment in the world. It's interesting that Germany could buy materials during the war only with gold. In extremis fiat money is accepted by nobody and gold is always accepted and is the ultimate means of payment and is perceived to be an element of stability in the currency and is the ultimate value of the currency. And that historically has always been the reason why governments hold gold."

Central banks, investors, and institutions alike all hold gold because there is no other asset or investment that has proven itself, and has stood the test of time, like gold.

So major gold sales from central banks seem unlikely. But there's still one major gold owner we haven't discussed in detail: the IMF. This organization owns over 100 million ounces of gold. What are the chances of the IMF dumping its gold on the market?

The answer's in the next chapter!

CHAPTER 9

Gold and the IMF

"December 3, 1997 was a critical date in Asian history. It was like so many of those dates in previous centuries in which foreign interlopers had first got to see the Korean king, threatened Japan with their black ships, and made their way in gunboats up the rivers of China. This time they had come in dark business suits...There were 17 of them from the IMF in Seoul; they stayed for a week to ten days, and they left having pried open a system that had refused to do what it had to do of its own free will...

"The IMF mission unleashed a storm of complexes, of ill-defined frights and fears and nightmares of goblins in the night. It was one thing to pay off a few debts, but another to sell out to the wicked Satan." [51]

Donald Kirk, correspondent for the International Herald Tribune, *on the IMF bailout of South Korea in 1997—which was bitterly opposed by many in that country.*

What is this mysterious entity called the IMF? Russia, Mexico, Brazil, Argentina, Thailand, Indonesia, Korea...these and other countries have been "bailed out" by this enigmatic, sometimes-reviled organization.

The IMF is sometimes said to be a paradox. Distributor of tens of billions of dollars in aid...yet often resented by the recipients of that aid. Powerful enough to rescue collapsing nations, even the eighth-largest economy in the world (Brazil)[52]...yet controlled by a mere 24 people. A central part of the international banking system, where billions of "dollars" exist as mere numbers in computer banks...yet also the world's third largest owner (after the U.S. and Germany) of true physical wealth, i.e. gold.

We saw in the last chapter how the IMF was formed: how the IMF was originally formed to mediate an international financial system based on gold. When that system ended, we also saw how the IMF was cut adrift and had to reconsider its purpose.

The IMF sold a substantial portion of its gold in the late 1970s, but it still holds over one hundred million ounces. Many commentators have been sour on gold investments, claiming that the IMF is likely to sell off these reserves in the near future and tank the gold market. Are they right?

We don't think so. Although nobody can predict the future, it seems very unlikely that the IMF will sell its gold. Its previous sales took place in a period when the global financial system had just been completely transformed...and like the central banks, the IMF was forced to re-evaluate its role. Although the IMF and central banks alike made anti-gold noises at the time, all have now realized the importance of keeping gold as a reserve asset.

We've already discussed how this process occurred for central banks. Now we'll see how it happened for the IMF...

* * * * * * *

> *"Gold remains at the heart of a collective belief in the credibility of an international economy...a sort of 'war chest', indispensable for a tomorrow whose needs we can only guess at."*
>
> *A former managing director of the IMF* [53]

The IMF currently has 184 member nations. Each country has a Governor, which sits on the IMF's Board of Governors. This Board meets once per year.

However, the day-to-day work of the IMF is done by its Executive Directors: a board of 24 members who each represent a nation or group of nations. Each director has a different amount of voting power, which is adjusted periodically: in 2001, the United States director held 17.16 percent of the votes, Japan's director was second with 6.16 percent, and Germany was third with 6.02 percent. Other directors include one representing 21 African countries (3.23 percent), one from Egypt representing 13 Arab nations (2.95 percent), and one from Brazil representing nine Latin American countries (2.47 percent).

The executive directors don't work in a vacuum. The G-7 countries (Canada, the U.S., the U.K., Germany, France, Italy, and Japan) together control almost half the votes, and so they exert a tremendous influence on the IMF in general. Nevertheless, the voices of the smaller nations are still heard.

As mentioned earlier, the IMF was originally constructed to manage a world economic system based on a dollar that represented a certain amount of gold. When the U.S. cut the dollar's link to gold, this created a dilemma for the IMF. The gold-based system that it was created to manage no longer existed. What role could there be for the IMF in this new world?

Plenty, as it turned out. The IMF now claims for itself the responsibility for "ensuring the stability of the international monetary and financial system...The Fund seeks to promote economic stability and prevent crises; to help resolve crises when they do occur; and to promote growth and alleviate poverty."

Perhaps the most visible work of the new IMF has been the "bailouts" of troubled nations. Mohsin Khan, the Director of the IMF Institute (which trains the Fund's newly-hired employees) explains it this way:

> "Consider the case of an individual. He's faced with a negative net worth—that is, his liabilities, his debts, are greater than his assets—and his income is less than his expenditures. He's spending more than he's making.
>
> "How can he do this? Because he's got credit—he can borrow. But now he's maxed out on his credit cards. No one will give him credit anymore.
>
> "The bank says to him, 'OK, we will bail you out. We will advance you some money. But now, everything you do has to be controlled by a financial planner. We can't allow you to keep spending the way you have, because you'll just run out of credit again. The financial planner is going to do two things: He's going to help you increase your income and help you control your spending. So that, in fact, you can only spend, beyond your income, to the extent we supply you with credit. We'll give you a loan of $10,000. The most you can overspend is that $10,000. And the financial planner is going to set targets for spending and help you earn more income, so you can pay the money back.
>
> "Furthermore, you are going to be watched very carefully. You're not going to get the $10,000 all at once. It's going to be spread out over a year. If you're living up to your commitments, you'll get the money. If not, we'll have to talk again.'"[54]

As in the analogy, IMF member nations who get into economic trouble can ask the Fund for help, and receive loans. However, the "financial planners" (i.e., mandates for economic reforms) that accompany the aid are usually very unwelcome by the recipient nation. This is perhaps stating the obvious—if they weren't unwelcome the country would have done them already on its own. The quote that began this chapter ("the wicked Satan"), shows the attitude of many people in South Korea during the IMF's bailout of that country in 1997.

Although no longer a core part of the system, gold still plays a key role in the IMF's operations. Quoting from the IMF's 1996 Annual Report:

- "As an undervalued asset held by the Fund, gold provides a fundamental strength to the Fund's balance sheet. Thus, any mobilization [i.e., sale] of the Fund's gold should avoid weakening the Fund's overall financial position.
- "The Fund should continue to hold a relatively large amount of gold among its assets, not only for prudential reasons, but also to meet unforeseen contingencies.
- "The Fund must take great care to avoid causing disruptions that would have an adverse impact on all gold holders and gold producers, as well as on the functioning of the gold market."

So the IMF recognizes that gold is an important part of its assets, for a variety of reasons. Let's discuss some of them.

First of all, the IMF realizes that "unforeseen contingencies" might require the stability of gold held as a reserve asset. The IMF recognizes that holding all of its assets in global currencies can be troublesome if all or many of the currencies were to rapidly devalue. Under these circumstances, the price of gold should rise, counteracting the loss of value in the currencies and providing stability to the Fund's assets.

In addition, one of the primary roles of the IMF is to help its member countries, especially the poorer ones. The World Gold Council has pointed out that many of these countries rely heavily on gold mining in their economies. "Ghana and Mali, for example, depend on gold for around one-third of their total export receipts. And this is by no means all, since mining companies build roads, power lines, schools, hospitals and whole communities. The development of a mining industry also increases government revenue through royalties and taxes, brings in foreign investment, technology and technical know-how, increases direct and indirect employment and helps the development of legal, financial and administrative infrastructure."[55]

Thus, anything which decreases the price of gold—such as open-market sales by large gold holders—hurts these countries. "Any actions taken by the Fund which might be detrimental to the gold price could therefore have an impact on the economies of a number of its members... at the extreme (and ironically), a fall in the gold price could be the last straw which forced a country to approach the Fund for resources."[56] In other words, if the Fund sells too much of its gold, some of its members might require bailouts—directly contradicting the IMF's mandate to maintain economic stability among its members.

Also, the IMF's gold is an "undervalued asset." Most of it is still kept on the books at the original valuation of 35 SDR/ounce (an SDR, or "special drawing right," was originally equivalent to one dollar. It's been redefined somewhat since then, but the point is that the IMF's gold is officially valued at a tiny fraction of today's true market price). This gives the IMF a strong backing to its assets that would be lost if the gold was to be sold and converted into depreciating currency.

In the past, the IMF has used this undervaluation to its advantage. In the late 1990's, the IMF needed funding for its Heavily Indebted Poor Countries (HIPC) program. Rather than sell gold outright to raise the capital, the IMF made an arrangement with Brazil and Mexico, which were about to make payments on their IMF loans. The IMF sold almost 13 million ounces of gold to these countries at full market price (thus making a substantial profit, since the gold was previously carried on the books at a much lower value).

The gold was then immediately returned to the IMF, and was accepted (at the same price) to fulfill the loan payments that were due. Thus, the gold was "sold" and returned without ever leaving the IMF repository—in effect, the IMF merely used these transactions to re-value some of its gold to the full market price. It was then able to use the difference to fund the HIPC initiative.

Importantly, the IMF was able to use its gold to fund a needed program, without selling any gold on the open market. Therefore, the price of gold was not affected. This transaction was unusual, but should the need arise to do it again, the IMF still has 90 million ounces of undervalued gold on its books.

Finally, the IMF cannot sell its gold without an 85% majority among its Executive Directors. As the United States holds over 17% of the voting power, the U.S. has an effective veto over any proposed gold sales. Since the U.S. is the largest holder of gold in the world, and has not sold any for over 20 years, it seems that IMF open-market gold sales are not likely to happen anytime soon.

* * * * * * *

"The elderly [moneychangers] waited patiently in the narrow alleys twisting off the eight-lane avenue leading past Midopa Department Store and the central post office. Occasionally they chatted with passersby, then delved into large handbags or scurried into a room behind a nearby restaurant. 'It's very difficult to gain much of a profit these days,' one of the women, peering into a

pocketbook containing rolls of $100 bills and 10,000-yen Japanese notes, told me. 'You never know which way the money is going.'" [57]

Donald Kirk, International Herald Tribune *correspondent, in South Korea when their currency imploded in 1997*

In the post-Bretton Woods world, the markets move at lightning speed. With no links to gold, currencies can now go into freefall in even the strongest economies—the Korean money changers quoted above found their own currency's value cut *in half* in little more than a month, sometimes dropping 10% a day (the maximum allowed before trading was suspended).

What's worse, one country plunging over a financial abyss can suck others down with it. The Koreans caught the "Asian contagion" in 1997 from Thailand, and then gave it Indonesia afterwards. Many credit the IMF bailouts with narrowly averting the spread of the contagion to the U.S. markets.

Previously, the dollar stood mostly above the turmoil in global markets, but its position has weakened. The strong-dollar policy of the recent past is no longer maintained: indeed, the dollar now faces an uncertain short-term future, as the war on terrorism is requiring massive expenditures. Long-term, the dollar will be under further pressure: the population is aging, and this means fewer productive workers at the same time that health care costs will increase. As we discussed in an earlier chapter, the dollar no longer enjoys the strength it had in the past.

Also, in today's world, currency speculators can exert tremendous influence on the markets, in spite of government efforts to the contrary. Witness George Soros' one-day profits of over $1 billion shorting the British pound—or his (partial) responsibility for a 12% crash in the Russian stock markets, when he made unfavorable comments about her economy.

In this environment, gold reserves are a crucial refuge. Indeed, during the Korean crisis we saw the government pleading with its citizenry to help save the country from disaster—by donating their gold. As long as the member nations of the IMF recognize this, the chances for large IMF gold sales seem slim.

In any case, over the last few decades the large holders of gold (i.e., the IMF and its member countries) are becoming less relevant to the market. Central banks and international organizations now hold less than 25% of the above-ground supply of gold, down from over 50% in the early 1960s.

Private owners, especially in Asia, own most of the world's gold, and they accumulate more every year. The *Economist* commented in January 1999: "The Indian lust for gold remains unabated...Gold jewelery [sic] is the only form of wealth that many women can claim as their own." At that time, the estimated amount of gold in India (9,000 tons) already exceeded the amount in Ft. Knox.[58] In addition, the recent deregulation of the gold market in China (which was already the world's third largest consumer of gold) is an exciting development.

Even if IMF gold sales were to happen, they might not even have a large impact. The IMF sold almost 25 million ounces from 1976 to 1980, and the U.S. Treasury sold about 17 million ounces of its own over the same period. Yet gold went up by hundreds of dollars. The Bank of England sold over half its reserves from 1999 to 2002: yet gold was up more than 20% by the end of *that* period.

Indeed, in the last chapter we saw that during the Gold Pool era, the efforts of Belgium, France, Italy, the Netherlands, Switzerland, Western Germany, the U.K., and the U.S. *combined* couldn't keep the price of gold down. When gold wants to rise...it *does.*

* * * * * * *

> *"There can be no other criterion, no other standard, than gold—gold that never changes, that can be shaped into ingots, bars, coins, that has no nationality, and that is eternally and universally accepted as the unalterable fiduciary value par excellence."*
>
> *General Charles de Gaulle, President of France, calling for a return to the gold standard, February 4, 1965*

We can summarize as follows:

- The IMF is the world's third-largest owner of gold.
- It has publicly stated that gold remains an important part of its assets. Indeed, in a world where the primary reserve currency (the dollar) is weakening, gold's importance as a reserve asset is only increasing.
- Selling its gold on the open market would risk a negative impact on the gold price, and this would violate some of the IMF's primary reasons for existing (i.e., helping poorer countries, some of which rely heavily on gold exports).
- If the IMF needed to use its gold to fund its programs, it can re-value it to full market price via off-market "sales" to members, for immediate return in loan repayments. The resulting "profit" is then available for use by the Fund. These transactions have no impact on the price of gold.

- The IMF cannot sell its gold without the express consent of the U.S., which has shown no inclination to sell any of its own. Even then, another 70% or so of the voting membership would have to approve.

So there seems to be little chance of a substantial, open-market sale by the IMF.

The World Gold Council has studied this issue extensively and said:

"There is very little prospect of the Fund's membership, as reflected in the Board of Executive Directors, agreeing to an outright sale of the Fund's gold. There are enough important members who hold gold in their national reserves—and see the good reasons for doing so—to prevent this happening."[59]

Nobody can predict with certainty the actions of the IMF, or any of its member nations. But the people who run these organizations are not foolhardy, and surely they know the lessons of history.

As Robert Mundell, the Nobel Laureate economist, predicted in 1997: "Gold will be part of the international monetary system in the twenty-first century."[60]

CHAPTER 10

Current Economic Trends and Gold

Terrorist experts believe that the United States will be attacked again on its own soil, and probably in the next few years. We have faced these dangers in the past and we have recovered. If we do have another attack, it would be very difficult to recover economically because our problems are larger and the tools we have to solve our economic problems are losing their effectiveness. We are very vulnerable to a serious financial crisis.

For investors, gold will be a major solution to help them through the next financial crisis.

> Gold has been and always will be the only true form of money. An objective look at history reveals that gold has been the only universally recognized store of wealth for the past 5000 years.

In times of war, revolution, economic disruptions, and social and political scandals, gold has continually proven itself as the most reliable store of wealth, as we have discussed many times in this book.

And right now, the United States is facing not only a long and protracted war against terrorism, but also growing economic problems that will have a major impact on the markets and gold. These include:

- Growing dependence on foreign oil and the potential for an oil shock
- The devolution of banking and the Federal Reserve Bank
- The proliferation of derivatives
- U.S. budget deficits
- Ballooning consumer, corporate, and government debt
- Inflation
- And the falling dollar.

Let's look at these in detail.

Oil Shocks and Higher Oil Prices

All the major world recessions of the last 30 years have had their roots in either disruptions in oil supplies or high oil prices. All of the great industrial economic powers of the world are very dependent on oil. The cheapest and most abundant sources of oil come from the Middle East, an unstable and unpredictable part of the world. Investors and businesses have learned to keep their eyes on oil prices and the Middle East, and specifically Saudi Arabia, the world's largest oil producer.

Overall, the Middle East possesses approximately 65% of the world's oil reserves. Saudi Arabia has approximately 261 billion barrels of oil, which is about 25% of the world's oil reserves. Industrial economies such as the U.S. are very dependent on Middle East oil, and this leaves the world vulnerable to the vicissitudes of Middle East oil production.

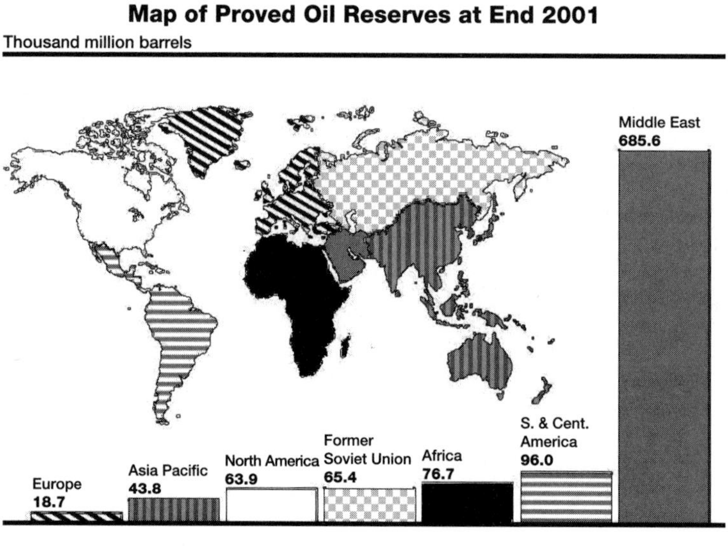

Map of Proved Oil Reserves at End 2001

Thousand million barrels

Europe 18.7
Asia Pacific 43.8
North America 63.9
Former Soviet Union 65.4
Africa 76.7
S. & Cent. America 96.0
Middle East 685.6

Figure 10.1: World's Proven Oil Reserves
(New York Times 2003 World Almanac, Edited by John W. Wright)

One of the most disturbing facts about the 9/11 hijackings was that 15 of the 19 participants were Saudis. Saudi Arabia is home of Islam's two holiest shrines, Mecca and Medina. The kingdom outlaws public worship of all religions except Islam.

There have been many news articles describing the fragile rule of the royal Saudi family and the ultra Islamist religious leaders who have enormous influence in Saudi society. Islamic leaders control much of the education of Saudi society. Young Saudis are taught a very extreme form of Islam and when they graduate from school, most have very few opportunities for employment. Young Saudis are ripe for recruitment for al-Qaeda and other Islamic extremist groups. Remember too that Osama Bin Laden is from Saudi Arabia.

The 2003 bombings in Saudi Arabia that killed 34 and injured around 200 are linked to al-Qaeda. Al-Qaeda and other Islamic terrorist groups are very interested in overthrowing the Saudi royal family and replacing them with an ultra-hard-line Islamic regime. Anti-terrorist experts report that al-Qaeda has wide support across all levels of Saudi society. Sympathizers give money, shelter, and other forms of aid.

Intelligence sources warn there will be other terrorist attacks in the Saudi kingdom including attacks on targets in the oil industry, especially pipelines that could be attacked easily but would cause major economic disruptions around the world. Other Saudi experts report growing hard-line fundamentalism that is spreading throughout the country. These conditions in Saudi Arabia are eerily similar to Iran in the late 1970s during the rise of the Revolutionary Iranian Islamic movement that eventually drove the Shah of Iran into exile.

You can see in the next chart what happened to oil prices when the Iranian revolution took place in 1979. The world saw an oil crisis, and the U.S. saw sky-high prices and long gas pump lines.

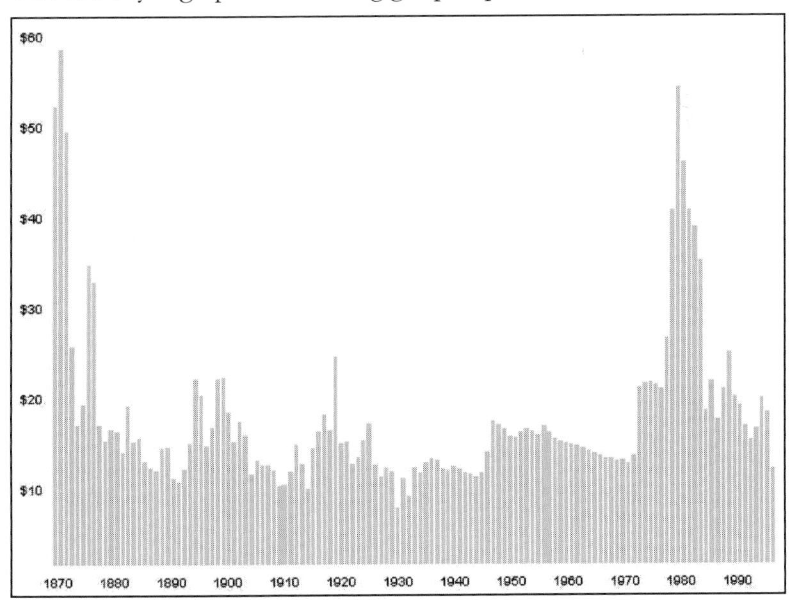

Figure 10.2: History of Oil Prices
(http://www.swiftenergy.com/SFY/Investor-Info/Industry-Outlook/2001/HistO&G/AR99OGme.htm)

The high oil prices of the late 70s and early 80s occurred during the Iranian revolution and the beginning of the Iran-Iraq War. The conditions in Iran in 1979 are similar to the conditions in Saudi Arabia today.

Some terrorist experts think certain strategic areas in the U.S. could be prime targets for terrorist organizations. For example, Houston has certain attractive advantages for al-Qaeda or other Islamic terrorist groups. The city has a fairly large Middle Eastern population to provide cover for any attackers. Also, several airlines provide direct flights from Houston to the Middle East.

The worst case scenario would be an attack along the Houston Ship Channel where a substantial portion of the U.S.'s refinery capacity is located. The entire area of refineries would create a huge explosion resulting in a cloud over Houston that would make the metro area unlivable for a significant period of time. This is an area where 4.25 million people live and work. The economic damage would be incalculable to the U.S. and world economy.

What makes matters even worse is that we are much more dependent on imported oil than we were 30 years ago. The graph below points out the increased vulnerability of the U.S. to oil disruptions because of the high amount of imported oil from the top five oil suppliers.

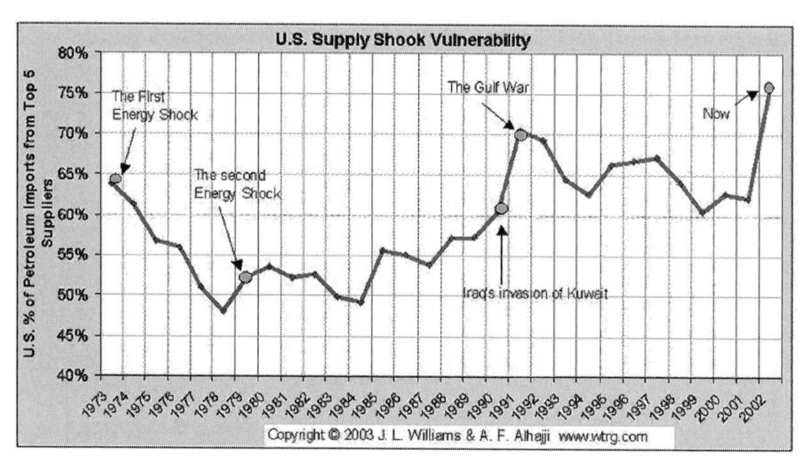

Figure 10.3: U.S. Percentage Imports for Top Five Suppliers

Oil shocks can devastate an economy. High oil prices mean high costs of transportation, which affects practically every business. Also, petroleum is used in making fertilizer, plastics, and many other things that are vital to our modern society. When crude oil goes up, the costs of many other items can skyrocket as well.

The next economic problem facing the United States that we want to consider is the devolution of banking and the Federal Reserve Bank.

The Changing Role of the Fed and Banks

Congress passed the Full Employment and Balanced Growth Act in 1978, and it mandated the Federal Reserve to (1) maintain price stability and (2) generate conditions to create full employment. The Fed has used three tools to achieve these goals:

1. Open-market operations
2. Reserve requirements of member banks
3. Changing key interest rates to member banks

The U.S. Federal Reserve Bank (the Fed) has had tremendous power over the economy for decades. They can expand the economy by lowering key rates and buying government securities in the market, and/or reduce reserve requirements to banks who would in turn lend capital to areas in the economy that had the most potential.

If the Fed wants to slow down the economy, they would reverse the process by raising key rates, selling government securities in the market and raising reserve requirements. During the down turn of every business cycle, bank sins would be exposed, especially through banks' bad lending practices. They would show up in real estate, Latin American loans, or in one of the worst cases, the savings and loans crisis.

The biggest changes that have occurred structurally in our economy are the increased practice of securitizing loans and the proliferation of derivatives. After decades of uncovering bad loans, banks and Wall Street decided to spread their credit risk by packaging their loans and selling them to insurance companies, institutional investors, and ordinary investors.

Theoretically this sounds good. Loans are spread out among many investors and the banking system is healthier and safer. Bankers and Wall Street point to the recent tech and telecom bubble to prove the supposed advantages of these schemes. The big banks and Wall Street maintain that if all the technology and telecom loans that were made in the 90s and the bankruptcies of WorldCom, Adelphia, and Global Crossing were on the banks' books, then we would have had severe problems for our banks and the banking system.

This is probably true. U.S. banks are healthier in the recession of the early 2000s than past recessions...but there are many problems that can and do occur with these new changes:

1. The Fed is losing control of the economy to the capital markets
2. Loans are based on marketability, not creditworthiness
3. Improper allocation of capital
4. Borrowing short term to lend long-term
5. Few people understand all the derivatives that are being created
6. Credit booms tend to lead to asset bubbles

It is becoming increasingly difficult for the Fed to achieve its mandate of full employment and price stability with its traditional tools. The Fed has dramatically pumped money into the economy since 1995 to avoid economic problems caused by:

- potential problems with Y2K
- the default by Russia on its debt
- the Long Term Capital Management crisis
- and the collapse of several Asian currencies.

The downside to keeping our economy going by increasing the money supply was the creation of asset bubbles, especially in the technology and telecom sectors, in the economy and the stock market. The Fed's tools of lowering rates and pumping money into the economy are not working effectively in the early 2000s, as the rates are close to zero, the economy is vulnerable, unemployment is rising, and potential bubbles are occurring in the bond and real estate markets. If we have another crisis, the Fed will not be able to lower rates to keep the economy going as they have done in the past.

We have talked about the potential problems from oil and from terrorist threats, but we want to also discuss another problem that has plagued our markets and the economy in the past and could plague us in the future: derivatives.

Derivatives

Derivatives are a very broad category of assets. They include convertible bonds, options, futures, and forward contracts. Derivative values are based on the performance of an underlying asset. Derivatives are used by hedge funds, institutions, companies, speculators, and individuals.

This area of investing can be very complex but we will keep it basic and brief and simply highlight past problems derivatives have created for our markets and the economy. Warren Buffett in his 2002 annual report warned of the potential problems our economy faces with derivatives, and he speaks from his own experience from operating a reinsurance company that Berkshire-Hathaway bought and the problems that company has faced with derivatives.

Briefly, the problems of derivatives that have appeared in the past and that may crop up in the future include leverage, liquidity, and pricing (mark to market).

The leverage involved in derivatives is substantial, and when it goes against the investor or institution it can wipe out capital and/or create maintenance calls that the holders can not meet.

Liquidity is the ability to buy and sell an asset quickly and in large volume without affecting the asset's price. Unfortunately liquidity can dry up in a market, forcing buyers or sellers to accept losses to exit a trade. This is like the "roach motel" where you can "check in" but you can't "check out."

The other problem that Buffett has pointed out is that many of these derivatives are on companies' books and their values are overstated. In addition the contra parties of some of the derivative instruments could be suspect in terms of fulfilling their contractual obligations. Part of the Enron mess was their use of derivatives and overstating the values of these instruments on their balance sheets.

The Stock Market Crash of 1987

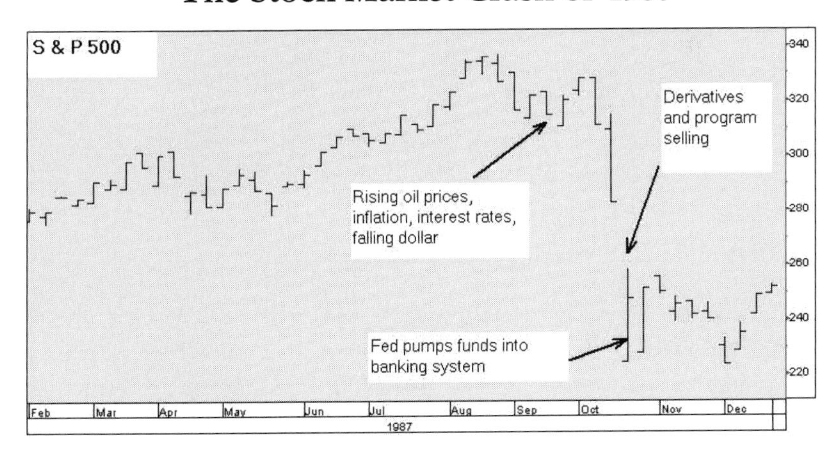

Figure 10.4: 1987 Stock Market Crash

The stock market crash of 1987 was one of those rare market meltdowns that were not followed by a recession. The Fed aggressively pumped money into the banking system, which lowered rates and probably prevented a recession. The financial press and analysts blamed the crash on many factors including: possible disruptions of oil due to the Iran-Iraq war, a falling dollar, and rising inflation and interest rates. These factors are the causes that led to the crash, but the selling became intense due to portfolio insurance, derivatives, and program selling.

In 1994, another meltdown occurred in 30-year treasury bonds. As we can see from the next chart, the yields for the 30 year treasuries in 1994 went from around 5¾% to above 8% in just one year. Yields and prices are inversely related. If you bought a bond paying 5% and rates go to 8% you would have to discount your bonds dramatically to attract investors to your lower paying bond. This is what happened to investors in 1994. Rates were reversing all year but the last straw was Orange County's financial collapse. Investments in derivatives created margin calls that forced Orange County, California into bankruptcy.

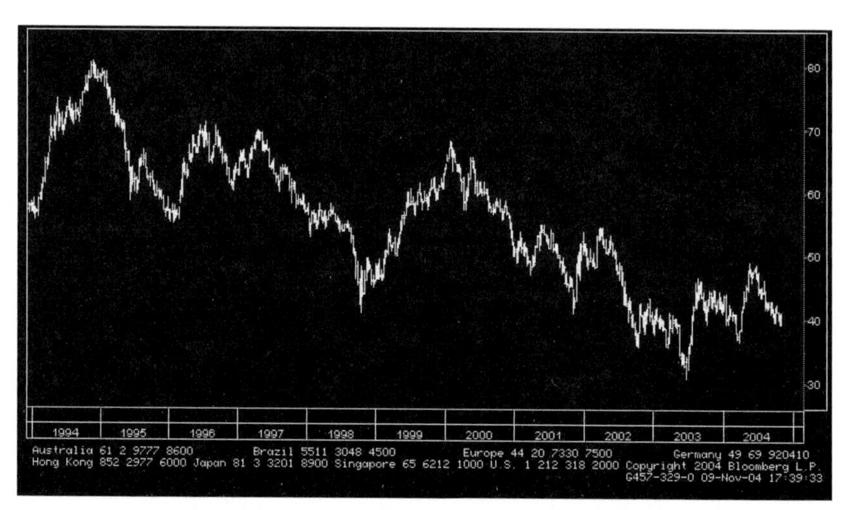

Figure 10.5: CBOE 30 Year Treasury Yield Index (Divide number on right scale by 10 to get yield.)

If yields reverse dramatically in the future (which they've started doing), bond prices could get whacked again. It wouldn't be surprising that another institution would be using the same derivative strategies and we would see rates spiking again, causing more problems for the economy.

1994 was a tough year for investors. The year started with the Fed raising key interest rates. Later the Mexican peso was devalued, and toward the end of the year Orange County filed for bankruptcy protection, the largest municipal filing ever. The stock market essentially went sideways in 1994, starting at the 470 level and ending at the 490 level:

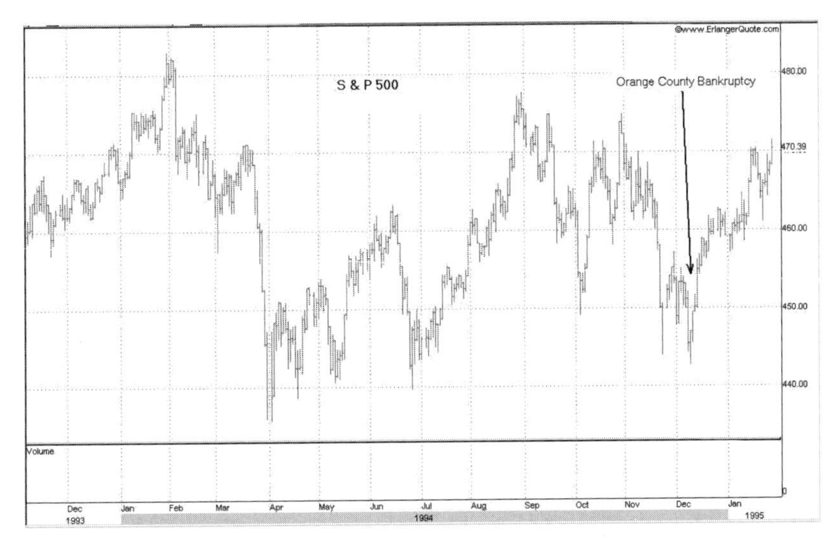

Figure 10.6: 1994 Impact of Orange County Bankruptcy on the S & P 500

The Asian contagion and Russia's default on its government bonds created enormous problems for the money management firm known as Long-Term Capital Management (LTCM), and for the Federal Reserve and major banks. The "elite" hedge fund had leveraged its equity of $4.8 billion into a $100 billion portfolio of derivatives. The fund was essentially short treasuries and long international debt securities. The Russian default caused investors to flee to treasuries and dump many international bonds:

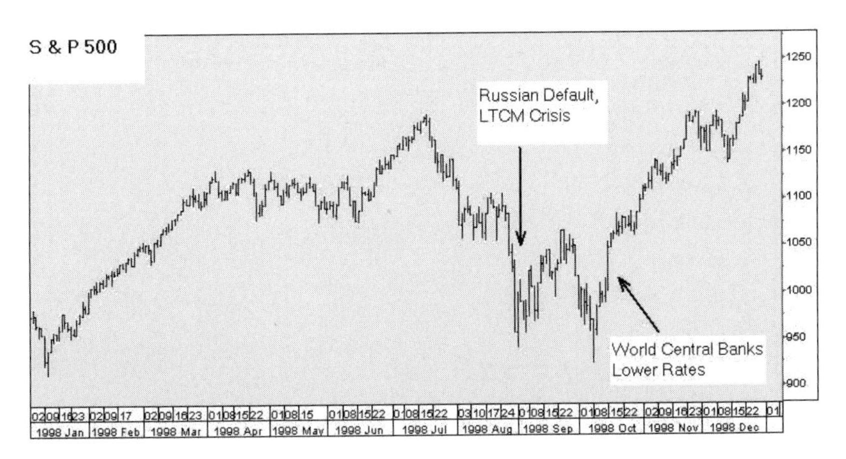

Figure 10.7: Russian Default and the Long Term Capital Management Crisis

LTCM was stuck with huge losses and it had to sell its huge positions to cover those losses, making the problem even worse. In October the Federal Reserve Bank of New York had to arrange with 15 major banks to negotiate a bail-out. Over the next few days central banks from around the world aggressively lowered rates and pumped up the money supply, and we were saved from a major financial crisis.

Bottom line: the Fed kept the United States out of recession for many years, but the tools of lowering rates and increasing the money supply are losing their effectiveness, and the capital markets keep creating bubbles.

Next on our laundry list of major economic problems are the U.S. budget deficits.

U.S. Budget Deficits

Debt can be a helpful financial tool but it is a double-edged sword. It can help finance needed long-term purchases and projects, aid in getting through tough economic times, establish a credit history and credibility, and it can even enhance investment returns if used judiciously. Easy credit can also enhance prosperity and stimulate an economy. But debt can also lead to severe financial problems if it is not managed properly. This seems to be the path that the United States is taking.

The official government debt of the U.S. is around $6.8 trillion, a large and incomprehensible sum. To put the debt in perspective, the share of each citizen's debt (population of U.S. is approximately 293 million) is approximately $23,800, an outrageous sum. The sum is not only enormous but the interest service of the debt alone makes the debt grow about $2.11 billion per day.

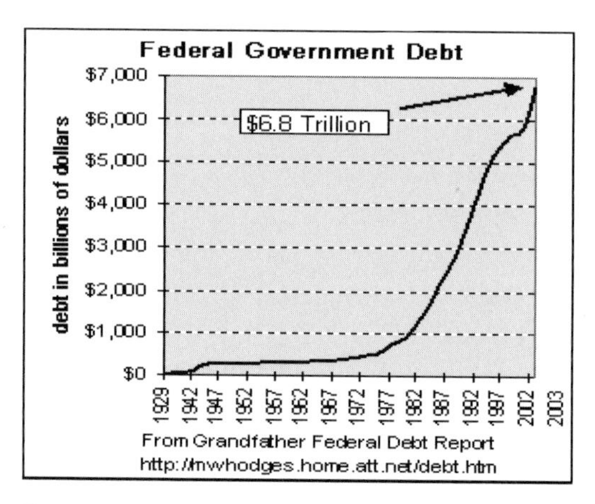

Figure 10.8: Federal Debt

Every single citizen should be extremely concerned about this debt—for several reasons:

1) Americans do not save enough to finance our debts (by buying government notes), so interest rates will have to be raised to attract foreign dollars to finance these debts. These huge debts will also cause all interest rates to rise as the government's financing needs will crowd out the credit markets and raise interest rates for corporations and individuals.
2) The weight of the debt and its interest payments will hinder the potential of the economy. A burden of debt reduces the potential of any economy, as it must use capital to pay principal and interest instead of buying goods and services.
3) This debt will create more problems in the economy if the economy slows down or faces any disruptions from potential geopolitical threats. In the worst case, the government could continue to print money to pay these debts, and this could lead to much higher inflation and a much lower standard of living.
4) The pace of the growth of the debt is quite alarming. We had mentioned earlier that the debt is growing because of accumulating interest, but there are other reasons for the growth in our debt. The chart below, "Federal Budget" shows how the U.S. budget went from a surplus of close to $300 billion in 2001 to a budget deficit of around $400 billion for 2003 with forecasts of $500 billion for 2004. These deficits will add to our growing debt problem.

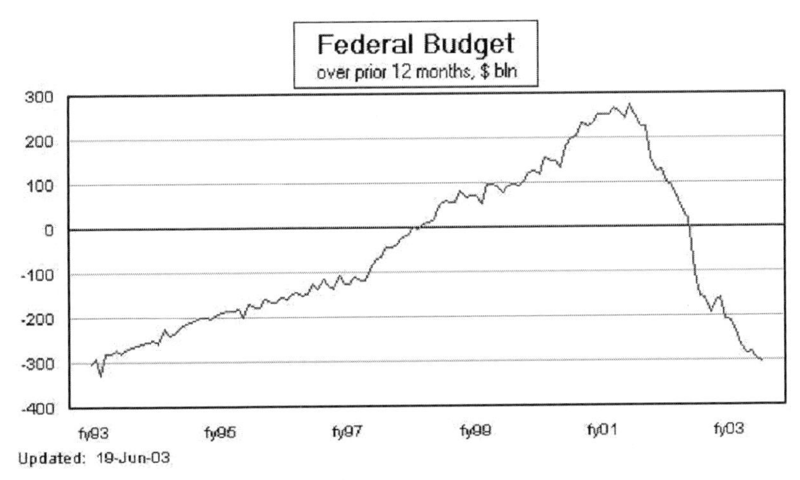

Figure 10.9: Federal Budget

The next chart, "Federal Receipts and Outlays," graphically illustrates the problem. We can see from the chart that spending has increased dramatically to fight the Afghanistan and Iraqi wars, the war on terrorism, and a new Homeland Security department and programs, while tax receipts were reduced because of tax cuts:

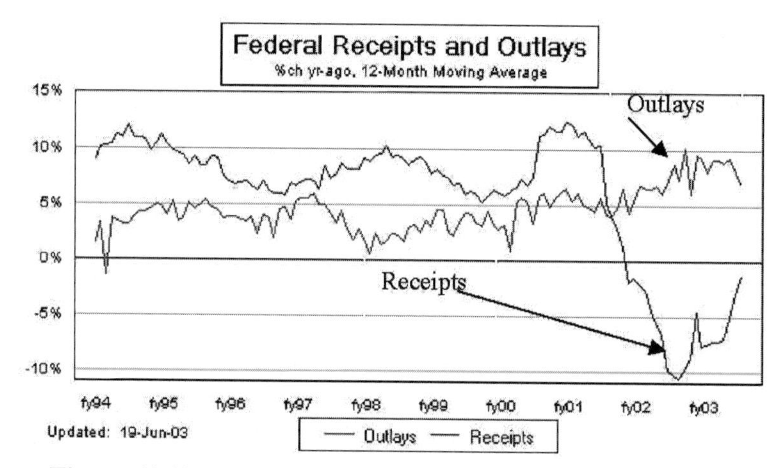

Figure 10.10: Federal Receipts and Outlays

Most economists agree that when our deficits go over 3% of our economy, the deficits become inflationary. This would be similar to having a $100,000 income but spending $103,000. This is fiscally irresponsible and financially unhealthy. It also simply cannot continue.

Some experts believe these debts do not fully reflect all of the government's obligations. Paul O'Neill, the former Secretary of the Treasury, ordered a study on U.S. government spending during the early years of the George W. Bush administration. The findings were very disturbing and were not published. These potential debts are a serious problem and warrant further discussion.

The study was finished about the same time that President Bush was pushing his tax cuts, and the financial press speculated that the study conflicted with Bush's tax proposal and would have given Bush's opponents ammunition to kill his proposals. The report highlights the mounting liabilities the government—and ultimately we the taxpayers—face in the future.

The forecasted budget deficit is $7 trillion for Social Security, and $36.6 trillion for Medicare.

The study is based on the work of Professor Larry Kotlikoff, Professor of Economics at Boston College, who developed what is called "generational accounting." This accounting takes into account future government liabilities in today's dollars. It would be similar to determining how much savings you would need to retire today. The answer for the United States is (conservatively) $44 trillion. That is 4 times the size of the current economy and about 6.5 times our current debt. Our country is not prepared to finance these liabilities.

The $44 trillion is based on everything the government expects to spend and what the government expects to earn far into the future. The difference between spending and income is discounted into today's dollars, which is $44 trillion. The lion's share of the spending is from Social Security and Medicare. What's more, many of the assumptions that are used in the projection are conservative. Part of the problem is that baby boomers will start retiring, thus reducing tax revenues while at the same time they will start applying for Social Security and Medicare. There will be fewer workers to pay taxes to support the increased government spending, especially as the first massive wave of baby boomers start retiring in about 5 years.

An article in the *Wall Street Journal* stated that baby boomers (the group that was born from 1943 to 1960) have the largest spending increases of healthcare. Self-centered baby boomers have higher incidences of obesity, diabetes and hypertension. The higher spending can also be traced to better medical technology, more consumer awareness and demand for powerful drugs and state-of-the art treatment, according to the *Journal.* If this trend continues, as it probably will, then the estimates above will be too conservative.

In 2003 health care already consumed 14% of our gross national product, and it's expected to continue to increase as the health needs of this large group increase.

As the baby boomers age, the per capita spending increases dramatically. From age 30 to age 70 there is nearly a five-fold increase. Multiply that times 77 million (the number of baby boomers), and you can understand why many economic experts are concerned.

There have been many proposals to address our ballooning debt, including:

- increasing federal taxation by 69%
- cut federal spending by 50% immediately
- increase immigration
- and delaying the retirement age.

All these suggestions would potentially solve the problems, but it is highly unlikely that these proposals would be implemented because they would be politically impossible to pass.

The reasons why the deficit study was not published, and the research and conclusions are not widely known, are why nothing will probably be done. No one wants to deal with the problems. Our massive debt is one of the biggest problems facing our nation. But the problem can't be fixed unless it's acknowledged. Experts have also determined that the longer we postpone addressing these liabilities, the larger the liabilities will become. If the government doesn't do anything the deficits get worse. By 2008 the deficit could reach $54 trillion. They also believe the Bush administration has made the situation worse by cutting taxes and increasing spending sharply, as we reviewed above.

Now let's take a look at total debt in the United States.

The Aggregate Debt Problem

We had mentioned in the last section that U.S. Government debt is a major problem, but if you add consumers and corporations to the government debt, the debt problem gets much worse.

In 1980 total aggregate borrowing of individuals, corporations and the U.S. government totaled $4 trillion. By 2002 that debt had grown to $31 trillion dollars, according to the Federal Reserve. This debt could lead us to stagflation, where interest rates will be forced up to attract foreign dollars to finance our debts, and the higher rates and crushing debt will keep our economy in stagnation. Stagflation also includes inflation (which we'll discuss in the next section). This debt does not give us much room for error, as any type of economic shock can be devastating to the U.S. economy.

The chart below illustrates the growing debt problem: debt as a percentage of GDP is greater than the ratio during the Depression.

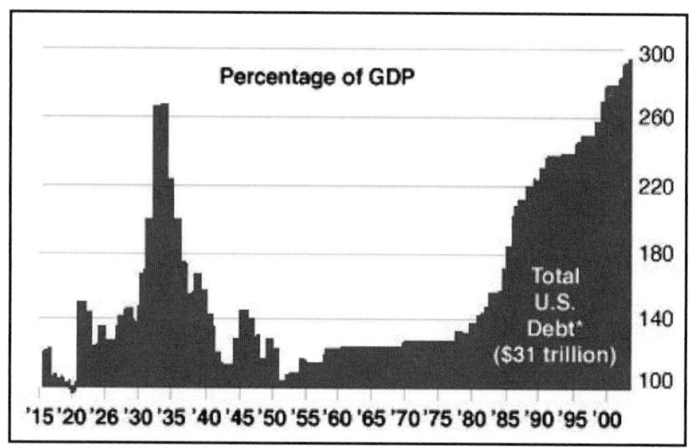

Figure 10.11: Total Consumer, Corporate, Government Debt as a Percent of GDP

The United States' financial condition is similar to a family living paycheck to paycheck, where they do not have an extra penny to spare or spend, all income is used to pay bills and pay interest and debt. If this family has an emergency, such as a major car or home repair or a serious illness in the family, this event could force the family into a much lower standard of living, forced asset sales, or even bankruptcy. What happens to this family can happen to our economy on a much larger scale.

The table below breaks down the sources of U.S. debt. What is very disturbing about this table is that a significant portion of the debt is from the government. We have close to an $11 trillion economy, and the government is approximately 18% of our economy, but it owes 26% of the total debt.

Source	**Debt (in Trillions)**
U.S. Govt.	$6.8
State & Local Govt.	$1.5
Mortgages	$5.8
Consumer Debt	$1.7
Corporate Debt	$5
Financial Institutions	$10
Approximate Total	**$31**

Figure 10.12: Sources of U.S. Debt

NOTE: The major theme of this book is that you need gold in your portfolio, but our advice is also to pay down your debt.

Now let's look at inflation.

Rising Inflation

In the last few years the world economy has seen trouble in a few high profile currencies like the Thai bhat, the Russian ruble, and the Iraqi dinar.

To combat economic hardship, central banks around the world have been lowering rates and increasing their money supplies. The impact of pumping huge amounts of money into an economy brings the problem of too many dollars chasing too few goods, i.e. inflation. Let's take a look at some disturbing trends we see in U.S. money growth and inflation.

We can see from the next chart the dramatic increase in the growth of the money supply since 1995. Many economists believe easy money is one of the main contributors to inflation.

M2 growth rate

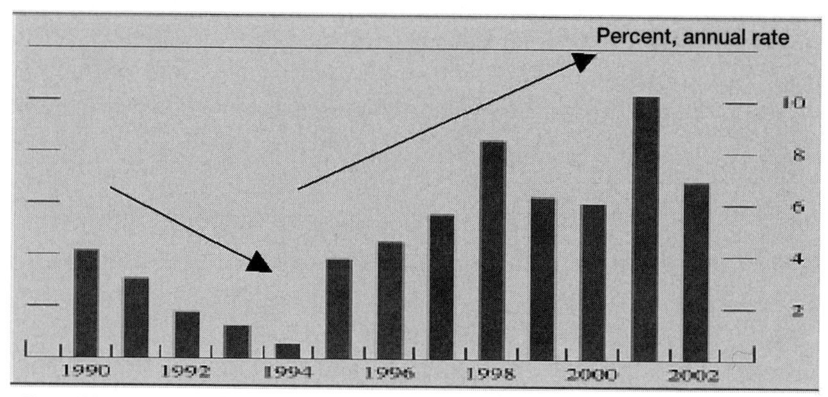

Note: M2 consists of currency, travelers checks, demand deposits, other checkable deposits, savings deposits (including money market deposit accounts), small denomination time deposits, and balances in retail money market funds.

Figure 10.13: Easy Money

In the chart below we can see the decline of CPI, the Consumer Price Index-All Items, and its recent reversal. The strongest price increases are in energy, food, housing, and commodity prices. Many economists are forecasting a slowing of home prices and sales but continued price increases for commodities and precious metals.

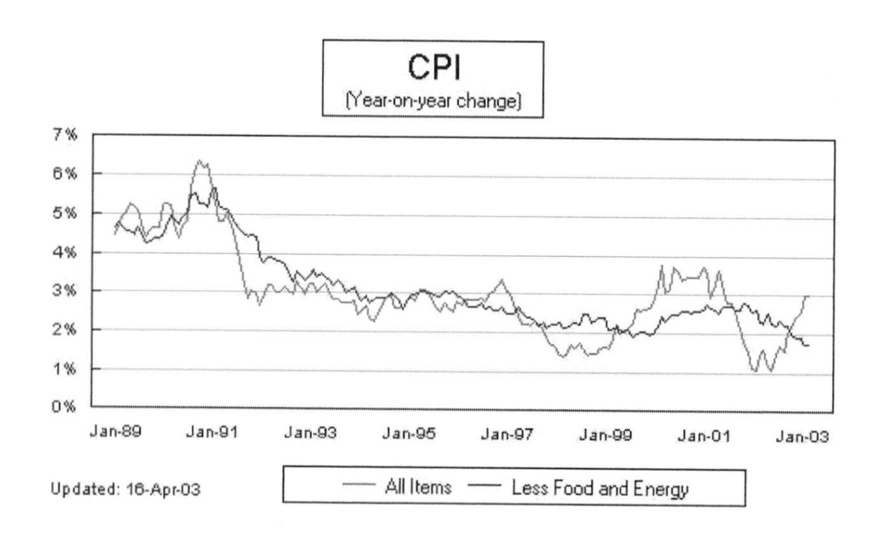

Figure 10.14: Trend of Inflation in the U.S.

You may be asking yourself, if all this money is being pumped into the economy, where is all that money going? For a start, we can see from the charts below that a lot of money in the U.S. economy is going to home sales, and we can see the inflation in the prices of homes.

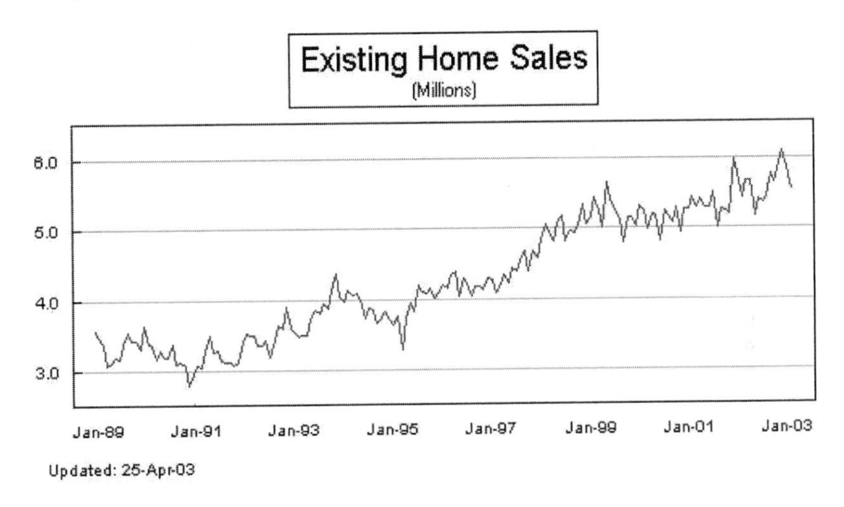

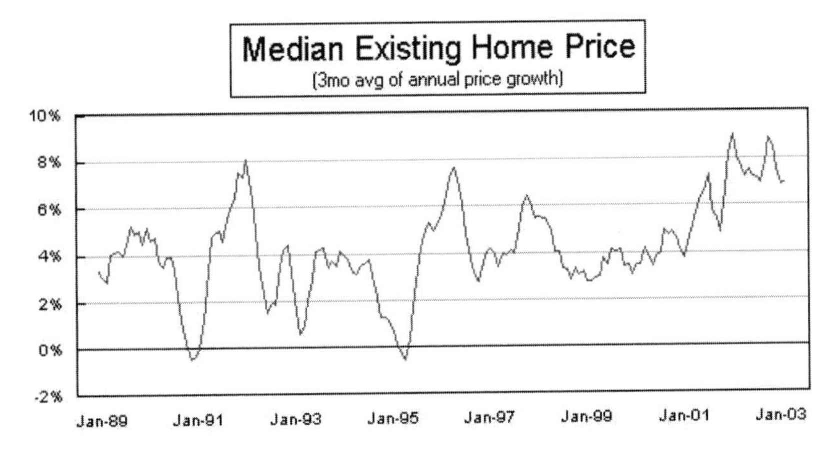

Figure 10.15 (A, B): Existing Home Sales and Prices

We also have inflation in bonds: higher prices and lower yields. Much of the money supply has moved into bonds. Investors have rushed into them in the last few years because of their perceived safety, the weakening economy, and the prospects of lower rates. But lately, rates have turned around and are starting to rise again. Bond investors will be hit hard if rates go up substantially.

We believe inflation will reverse, and that can be very damaging to your bond holdings. As inflation and interest rates rise, if you need to sell your bonds you will have to lower the price of your bonds to compete with higher yielding bonds. If interest rates move only one percentage point, from 5% to 6%, you could lose up to 20% on your principal if you own very long-term, non-callable bonds. If you own bonds, make sure you protect yourself. The actions you should consider are taking profits, reducing your holdings, hedging with futures, or holding your bonds until maturity.

One of the concerns that stock market experts have is that with all this cheap money available, speculators have been borrowing short-term to speculate in the markets. This can exacerbate market movements if rates do reverse. Borrowing short-term to speculate is a profitable strategy as long as prices are rising and rates remain low, but once rates start to rise then borrowing costs start to go up and asset prices start to fall. These speculators are forced to sell, creating even more selling in the bond and stock markets.

The Fed has stated that they are willing to keep interest rates low for a long time, but as we mentioned the Fed has lost control over the economy and most interest rates. The chart below demonstrates how rates have bottomed and have already started to reverse, despite what the Fed says:

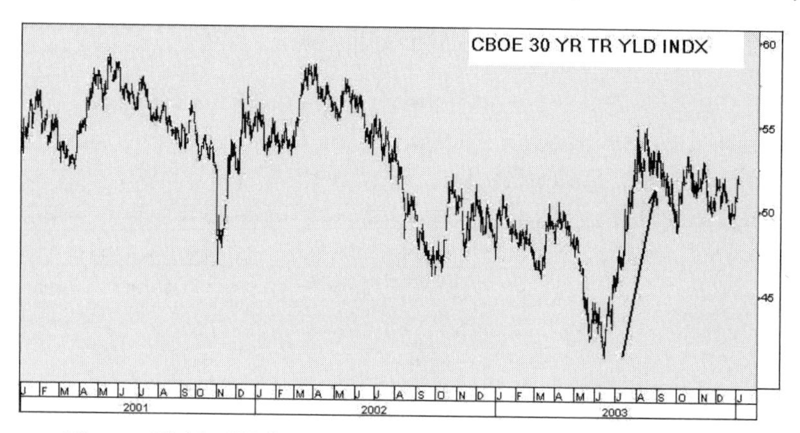

Figure 10.16: CBOE 30 Year Treasury Yield Index (Divide number on right scale by 10 to get yield.)

We can see from the chart above that long-term rates have risen about 1% since the finding a bottom in early 2003, despite the Fed's statements that they will keep interest rates low.

There is over $5 trillion sitting in money market type accounts. We don't think that investors really know what a bad deal this is. The real returns on money funds are negative. Money market yields are around 1% minus inflation of 1.5% and minus taxes leaves you with a negative return. It would be wise to move some of your money from money markets accounts and bonds into gold. Gold can protect you from imminent inflation.

The price of gold bottomed in the first quarter of 2001, and has been slowly climbing, as we can see from the following chart:

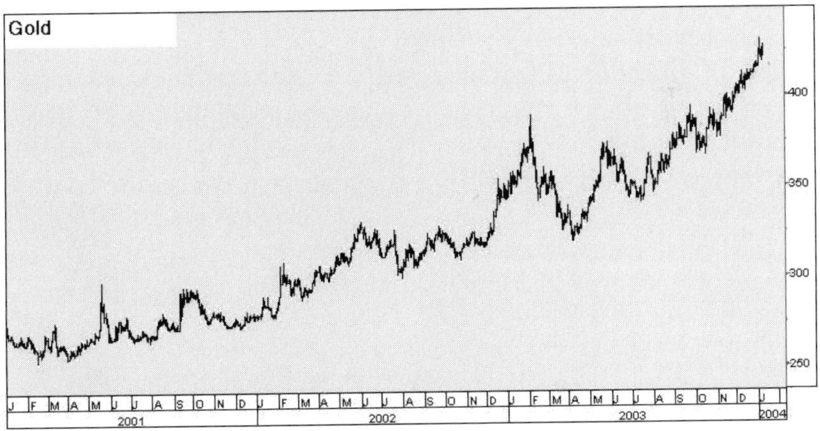

Figure 10.17: Price of Gold After Bottom in 2001

There are a few causes offered to explain the increase, but many believe gold is anticipating inflation. Alan Greenspan believes in the predictive value of inflation for gold. In the 90s he spoke before a Congressional hearing and he cited gold as a "measure which has shown a fairly consistent lead on inflation." People want to hold real assets rather than financial assets when they think money's purchasing power will drop, he explained. As a gauge of inflationary expectations, gold is "better than commodity prices or a lot of other things," he added.

And finally, on our laundry list of economic problems, is the falling dollar.

Falling Dollar

The chart below shows us how much of a decline there's been in the dollar just since early 2002:

Falling Dollar

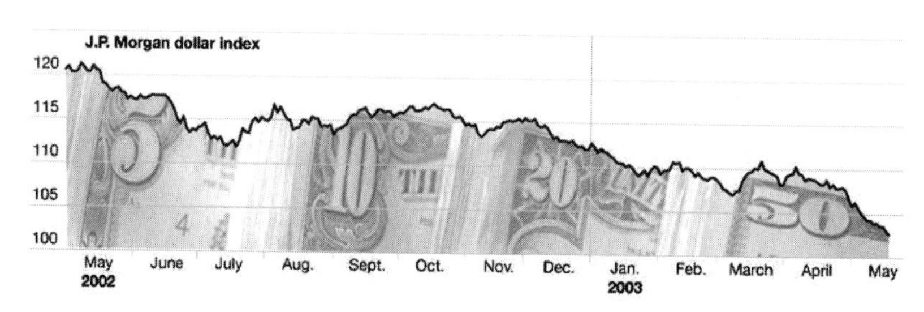

Figure 10.18: Dollar Decline Since 2002

There are many causes cited for the rapid decline of the U.S. dollar that started in 2002. For many years, foreign investors and businesses moved money into the U.S. markets and economy, based on the expectations of a healthy growing economy. A strong dollar is evidence of confidence in our economy and investment opportunities from our trading partners and investors.

The confidence in the dollar has eroded. The view of the U.S. as a place to do business and invest has changed for the following reasons:

- Stock market bubble burst
- 9/11, higher risk to do business in the U.S.
- Fed has lowered interest rates 12 times to historic lows, making dollar denominated paper less worthwhile to hold
- Corporate scandals, negligent auditing firms, and government regulators who were asleep at the wheel
- War on terrorism, including wars in Afghanistan and Iraq
- Twin deficits, U.S. budget and trade balance (Imports vs. Exports).

We have seen a net outflow of money from foreign investors, causing downward pressure on the dollar and the U.S. stock markets. Long-term, the dollar has fallen 34.4% since 1986.

A Lower Dollar Is Inflationary

Let's discuss this.

For example, if you had converted a U.S. dollar into a Euro last year you would have received one Euro. Today, if the dollar is weaker and you convert one dollar into a Euro, you would need $1.25 to buy one Euro. A weaker dollar makes foreign goods more expensive and reduces foreign competition, leading to higher domestic prices.

What normally can happen is domestic suppliers and manufacturers know foreign goods are more expensive, and they now can raise their prices, potentially creating inflation. That's great news if you are a producer of goods in the U.S., but it is bad news if you are a consumer of goods. And inflation is even worse news if you are a bondholder.

If you're a holder of Euros or any other major currency, a dollar devaluation is good news. Let's say you were holding that Euro you bought last year for one dollar. Now you could exchange your Euro for $1.25. A lower dollar makes U.S. goods and services, including dollar denominated gold bullion, attractive to foreign investors. Foreign buyers in this case are getting a 25% discount on U.S. goods and services.

Most other countries are wary of their countries' currency stability as they have seen their currency devalue and destroy their savings and purchasing power. This is especially true of large importing countries. These countries know the value of gold as a store of value. U.S. imports represent only about 12% of our GDP, so a devaluation here is not as damaging. If the dollar continues to drop more than it already has, then we will learn the hard way the impact of a currency devaluation.

We think the weak dollar will be another impetus for inflation, and foreign investors will push up the prices of dollar-denominated gold investments. Also, if the dollar gets too low, the Fed will be forced to raise rates to attract foreign money to finance our twin deficits, bad news for bond and stock investors.

Another important fact to remember is that historically, when citizens lose confidence in their country's currency, they seek gold as a safe haven.

NOTE: It seems that the time-frame investors watch has become very small. Every little blip of news about the dollar or interest rates or oil prices or jobs is enough to send the markets shooting higher or dropping like a stone. The government also, for political reasons, wants you to be encouraged by short-term news. For both Democrats and Republicans, short-term news helps push their positions.

It's important to remember that the trends we have been discussing here are long-term trends. They may not move in a straight line—there may be short-term rises or falls in the debt, the dollar, inflation, oil prices, the price of gold, or any single part of the picture.

But long-term, these are powerful forces that continue to build. Historically, they cannot be ignored.

Part IV

Investing in Gold

CHAPTER 11

Direct Investments in Gold

In the last chapter we discussed the many reasons why gold has been and will continue to be in a bull market:

1. Terrorism fears and geopolitical risks: Since 9/11 and the Gulf War II investors have been moving their money into gold.
2. Inflation: As we discussed in the last chapter, gold is one of the best indicators of inflation. Investors are taking their cues from gold and jumping on board. Many economists believe the falling dollar will lead to inflation. As we discussed, a cheap dollar can cause prices to rise in the U.S. as the weaker dollar makes foreign imports more expensive.
3. Falling dollar: Gold is priced in dollars, so a weak dollar makes gold more affordable for most foreign buyers with their stronger currencies.
4. Ballooning deficits: The 2004 budget showed spending for the U.S. government at $2.4 trillion dollars, with a projected $521 billion deficit.

Investors need to protect the purchasing power of their investment dollar and hedge against the risks stated above.

The gold bull market has started and will continue. The respected investment publication *Barron's* interviewed James Turk, publisher of the *Freemarket Gold and Money Report.* He believes gold can reach as high as $8,000 an ounce. Mr. Turk's forecast is based on his belief that there will be a shift in valuations and investing from stocks to gold, similar to what happened in the 1970s.

The 2000s and 1970s are eerily similar: tensions in the Middle East, poor fiscal and monetary policies, economic uncertainty and risk. In the 1970s the Dow industrials traded between 600 and 1000. Toward the end of the 70s, gold and the Dow crossed at around 800. Mr. Turk believes the Dow and gold will cross again at 8000 at some point. Since the 1970s, the Dow has gone up by a factor of ten. But gold has not yet multiplied, despite its wildly bullish fundamentals—so Mr. Turk believes that it too will go up by a factor of ten from its $800 level of the late 1970s. Just as gold and the Dow crossed at 800 they will cross again at some point in the next few years at 8000.

This is certainly the high side of forecasts. But most predictions for gold are in the $800 to $1,000 range, surpassing the $800 high for gold in 1980.

Long-Term Gold Trends

The following chart is a long-term chart of gold prices since the mid-1970s:

Figure 11.1: Long-Term Chart of Gold

Some important highlights of the chart are:

1. Gold reached a high of $800 in 1980
2. Gold finds support in the $250 to $300 area that lasts for over 20 years.
3. The low for gold is established between 1999 and 2001 in the $250 area.
4. Gold breaks out of its bear market decisively and enters a bull market as it breaks above 300 in 2003.

If you believe in the evidence we've presented in this book—that holding gold is an absolute necessity for financial protection—then we've now come to the action part of this book...

How to Invest in Gold

The best thing about investing in gold is that there are so many choices. The worst thing about investing in gold is that you have so many choices.

In this chapter we will discuss the best way to invest in gold—through direct investments such as bullion, coins, and rare coins. The next chapter we will discuss the many ways to invest in gold indirectly through mutual funds, closed-end funds, and gold stocks.

Directly investing in gold is the best way to participate in the price action of gold for several reasons:

1. Most direct gold investments will move with the price of gold but some gold stocks may not, especially when the markets are extremely volatile. Gold stocks sometimes behave as stocks instead of gold investments.
2. If there is a crisis, trading gold for goods and services would be easier to trade than gold securities.
3. It's easier to invest in gold directly.
4. There are many choices in terms of size, quality and cost. There is a direct gold investment to fit any budget.

There are literally dozens of ways to invest in gold directly because there are different forms of physical gold bullion on the market. There is an assortment of coins and bars in all sorts of weights, shapes, and sizes, produced by dozens of countries and private mints around the world. In this chapter we'll discuss some of the most popular forms of bullion coins to give you an idea what's available on the market.

Gold Bars

The gold bars that you see in the movies are mostly traded on the major world commodity exchanges, and are used by the world's central banks and major institutions.

Private investors buying less than 1000 ounces of gold should steer clear of these 100-1000 ounce gold bars. We strongly recommend NEVER buying smaller gold bars like 1 ounce or less weighted gold bars produced by private mints or refiners.

First, small investors who buy one or two 100-ounce bars lose the ability to sell their gold in intelligent increments. If gold doubles as many experts suspect it will, it may make sense to sell 25% or 50% of an investor's gold holdings. 100-ounce bars make that very difficult. In addition, 100-ounce bars are only regularly traded on exchanges. Most gold dealers, coin dealers, and gold brokers don't trade 100-ounce bars and will discount a bar that large by 5% to 7%.

Secondly, the marketplace is dominated by bullion coins. The vast majority of rare coin and bullion dealers do 99% of their trading in coin form. Gold bars sell for a 3% to 10% premium over the spot price of gold, and when you sell them you receive 3% to 10% under the spot price of gold. That means if you bought and sold, and prices remained the same, you could lose as much as 20% of your invested capital.

It's important here to make a distinction between bullion coins and numismatic coins. A bullion coin's value is derived solely from the content of its gold and is normally sold at a small premium above the market price for gold. A numismatic coin derives its value from its rarity, historical, and aesthetic qualities. Numismatic coins commonly sell for thousands of dollars, and can sell for tens of thousands, hundreds of thousands, and even millions of dollars.

Private Mint Gold Coins

You must also be cautious about private mint gold coins. Many refiners and private mints around the world produce 1 ounce to 1/10th ounce gold coins and offer them for sale as "bullion" alternatives. They tout a lower cost than more commonly traded gold bullion coins produced by the governments of United States, Canada, South Africa, and Australia, or the coins are sold based on the uniqueness of their design. Private mints coin their gold bullion with images of everything from sporting events to Elvis Presley. They also sell for large premiums above the price of gold and sell at a discount to their intrinsic gold value because they are not widely bought and sold by dealers, and therefore dealers will discount the coins when (and if) they buy them.

So, what gold bullion coins can you buy safely for a modest premium? Stick with the five most commonly traded gold bullion coins in the world.

South African Krugerrand

Back in the 1970s the most famous gold bullion unit was the Krugerrand from South Africa. The coins contain one ounce of gold and just enough copper to allow the coin to be struck. So the net weight of the coin is actually more than one ounce. They dominated trading in the last gold bull market and are still traded today. The South African government produces small weight coins in addition to the one ounce standard. They have:

- No currency value
- Gross Weight: 33.933 Grams (1.0909 troy ounces)
- Fineness: .916 or 22 karats
- Diameter: 34mm
- And are also available in ½, ¼, 1/10 ounce coins

Canadian Maple Leaf

The popularity of the South African Krugerrand prompted the Canadian government to mint the Canadian Maple Leaf in 1979. The coin was an instant success thanks to a clever advertising angle that touted the maple leaf as the first solid 24-Karat gold bullion coin. While that is true, the fact remains that Canadian and South African coins both contain a full 1 troy ounce of gold. Few people realize the Canadian Maple Leaf actually has a face value of $50 Canadian Dollars, far less of course than the value of the gold bullion.

- Face Value: $50 Canadian
- Gross Weight: 33.1033 Grams (1 troy ounce)
- Fineness: .999 or 24 Karats
- Diameter: 30mm
- Also available in ½, ¼, 1/10 ounce coins

Australian Kangaroo

The "ROO," as it's commonly referred to, is minted by the Australian Perth Mint and is actually the second bullion coin produced by Australia. The first was called the "Nugget coin" and the Kangaroo replaced it.

- Face Value: $100 Australian
- Gross Weight: 31.1033 Grams (1 troy ounce)
- Fineness: .999 or 24 Karats
- Diameter: 32.10mm
- Also available in ½, ¼, 1/10 ounce coins

Chinese Panda

One of the most popular gold bullion coins in the world is the China Panda, which was first introduced in 1982. The 1/20 oz was introduced in 1983. Throughout the years, the China Mint has kept the same Panda design but has frequently changed the position of this Panda on its coins.

- Face Value: 500 Rinimbi
- Gross Weight: 31.103 Grams (1 troy ounce)
- Fineness: .999 or 24 Karats
- Diameter: 32.05 mm
- Also available in ½, ¼, 1/10, 1/20 ounce coins

American Eagle

The American Gold Eagle is now by far the most popular gold bullion coin in the world. Authorized by Congress in 1985 and first minted in

1986, American Eagles are minted in 22-karat which was the standard established for circulating U.S. gold back in 1796. In fact, the 22-Karat standard has been the worldwide standard for circulating gold coinage for over 350 years!

- Face Value: $50
- Gross Weight: 39.33 grams (1.0910 troy ounces)
- Fineness: .916 or 22 Karats
- Diameter: 32.7 mm
- Fine Gold Content: 31.103 grams (1 troy ounce)
- Also available in ½, ¼, 1/10, 1/20 ounce coins

Dos and Don'ts of Buying Gold Coins

Don't buy bullion coins that have any rim nicks, scratches, abrasions, chips, and dents or appear to be discolored in any way. NEVER! Any knowledgeable buyer will discount coins that have even the slightest damage.

Steer clear of any coins that have carbon or copper spots. Some gold bullion coins, even ones that are in 100% absolutely perfect condition, will have tiny spots visible to the naked eye without magnification. These are natural and caused by the inclusion of copper in the gold to make the planchets (on which the coins are struck) more durable. Despite the fact that these spots are natural to gold coins, they are undesirable and dealers will buy and sell them at a small discount. So, make sure when buying gold bullion coins you insist on "no spots." Keep in mind a spot is only a problem if you can see it with the naked eye. If you have to use a magnifying glass to see a spot, it is not a problem.

Don't buy "rare date" bullion coins. A bullion coin is a bullion coin. Don't be fooled. The least expensive way to purchase the 1 oz coin is to specify "common date." Common date means the bullion dealer can send you any date bullion coins of the type you desire in Gem condition. If you order a specific date, for example 1996, it will cost more than the common date.

NEVER buy or sell gold bullion strictly on the basis of the best price. Saving a few dollars when buying or selling isn't worth the risk of using an unknown dealer.

Know your dealer: do some background checking. How long have they been in business? Check with the Better Business Bureau. Are you dealing with a nameless clerk, or a principal in the firm that you can also check up on? You'd be amazed how many "colorful" people are out there waiting to steal your money.

Always take immediate delivery of your gold coins. NEVER store your gold coins in a dealer's vault. I've seen people lose every penny "trusting" a dealer. Take the time and get a safety deposit box at your bank and take charge of the storage. When buying bullion it's important to get your gold as quickly as possible. Checks need several days to clear, money orders less, bank wires are immediate, and you can always insist on next day shipment when you send a bank wire.

Rare coins tend to appreciate better than bullion gold coins. For example, Gem Uncirculated U.S. gold coins struck from 1796 through 1933 have consistently outperformed every bullion alternative for the past 33 years—without exception.

A "Gem Uncirculated" U.S. gold coin is a coin that remains in the original state of preservation—these quality coins look as they did on the day they were struck.

A gem quality coin may have a minor bag mark or abrasion but for the most part a gem coin's surfaces remain in a near-perfect state.

Rare coins are graded much like diamonds. Rare coin grading uses a scale of 1 through 70. A coin graded 70 is considered perfect. The designation "MS" stands for mint state. Coins in Mint State condition are graded MS-60 through MS-70.

It's virtually impossible to find an MS-70 gold coin from 1796 through 1933. Collectors prize Choice Uncirculated coins graded MS-64 and MS-65, and lust after Gem Uncirculated coins graded MS-66, MS-67, MS-68 and MS-69.

Gem quality gold coins from 1796 to 1933 are truly rare since few survive. To understand this rarity, you have to keep in mind that these coins were minted as spending money. Precious few collectors had the disposable income to put aside gold coins as collectables in the 19th century or in the early part of the 20th century, much less through the Great Depression of the 1930s!

Because of their scarcity, gem quality U.S. gold coins offer a genuine and unique one-way leverage that literally no other gold investment offers.

As the price of gold moved steadily higher during the 1970s the price of Gem Uncirculated rare U.S. gold coins rose even faster, sometimes outpacing the rise in bullion by as much as 2:1! For example, a Gem Uncirculated MS-66 $20 gold Saint Gaudens in 1974 went from $175 to trading for over $2,000 in 1980 at the peak of gold's rise. While gold bullion went from $145 to $850 an ounce, that Gem Uncirculated $20 Saint Gaudens enjoyed more than 2 to 1 advantage in price appreciation!

This amazing leverage became even more spectacular when gold peaked and entered what became a 22-year bear market. While the price of gold dropped, many rare Gem Uncirculated gold coins from 1796-1933 actually continued to increase in value.

In fact, if you had moved from gold bullion in 1980 to really gem rare U.S. gold and held the coins for the past 22 years, you would have actually made a substantial amount of money. There are of course exceptions to this rule. But as whole the Gem Uncirculated rare U.S. Gold coins have consistently been big winners over bullion gold year in and year out. Here are a couple of typical examples...

A Gem Uncirculated MS-66 1916-S $20 Saint Gaudens $20 gold coin in 1980 at the top of the gold market would have cost you no more than $2,000. Today, 23 years later that very coin now trades wholesale for $5,750. A gain of 187% while the price of gold fell from $850 in 1980 to just $342 in 2003!

A Gem Uncirculated MS-66 1929 $2.50 Indian gold coin in 1980 would have cost about $2,000 in 1980. Today, that same coin would trade wholesale between dealers for as much as $8,500. A gain of 325% while gold has fallen over 59%!

This unique one-way leverage is the result of the growing numismatic interest in these coins, which supersedes the effects of a bearish gold market. This is true in large part because of the U.S. government marketing efforts as mentioned in the next examples.

In the past 22 years, the number of rare coin collectors coming into the market has vastly outnumbered the supply of these rare gold coins because of the marketing efforts of the U.S. Mint.

Each year the U.S. Mint produces, promotes, and sells hundreds of thousands of U.S. Mint and proof sets, as well as a seemingly never-ending series of modern commemorative, half, dollar and $5 gold coins.

The current U.S. Commemorative Quarter program is another amazing promotion program used by the Mint to generate new coin collectors every day.

It's been estimated by industry experts that the U.S. Mint brings over half a million new rare coin collectors into the hobby every year. Even if a small number of these new collectors decide to collect truly rare coins like 1796-1933 Gem Uncirculated U.S. gold coins, the price of these coins must rise substantially regardless of whether the price of gold goes up or down. Again, this one-way leverage is amazing.

To put this in perspective: only 96 Gem Uncirculated 1916-S $20 Gold coins survive in Gem Uncirculated MS-66 condition. And only 2

1929 $2.50 Gold Indians survive in Gem Uncirculated MS-66 condition. If only 1% of these new collectors entering the market decide to collect these Gem Uncirculated gold coins, you'll have 5,000 new buyers chasing these few coins. In the last 33 years this has produced a steady rise in the value of these coins. Can you imagine what the price gains for these coins will be in a new bull gold market? It's going to be explosive!

What to Know Before You Buy A Single Rare U.S. Gold Coin

There are two independent and reliable grading and authentication services in the rare coin market. They are NGC "Numismatic Guarantee Corporation" (www.ngccoin.com) and PCGS "Professional Coin Grading Service" (www.pcgs.com). NEVER buy a rare coin that hasn't been graded and certified as authentic by one of these two services. NEVER! Both services ultrasonically seal the coins they grade and certify in airtight plastic holders—these protect the coin and make storage and resale easy.

There are lots of rare coin grading services that attempt to imitate NGC and PCGS in the marketplace, but they are NOT reliable. In my experience, rare dealers that try to sell coins that are not graded by either NGC or PCGS are less than ethical.

Both NGC and PCGS grade their coins based on the Sheldon scale (from 1-70) that we described at the beginning of this chapter. We recommend both Choice and Gem Uncirculated MS-66 condition coins most often—they're beautiful and rare, with excellent investment potential, but are more affordable than the higher grades. However, many of the coins we love and think offer the best investment potential simply are unavailable in Gem MS-66 condition. Sometimes the best possible coins are only available in Choice Uncirculated condition MS-65.

Building a Gem Uncirculated U.S. Gold Type Set

One of the best ways to position yourself and diversify your investment portfolio is to assemble a Gem Uncirculated 20th Century (pre-1933) U.S. Gold Type Set. Depending on your budget, there are several possible sets that can be assembled. The 8-piece type set is the most commonly assembled set by investors, and it includes an example of all 8 U.S. gold coins minted by the United States for circulation from 1900 through 1933.

An 8-piece 20th Century gold set includes two examples of the $2 ½, $5 and $10 gold coins: one each of the Liberty and Indian Types, as well as one example of each of the Liberty and Saint Gaudens $20 gold coins.

Assembling a Gem Uncirculated MS-66 8-piece gold set graded and certified by either of the two most respected independent grading services (NGC or PCGS) is in our opinion the single best gold physical investment an investor can make. The coins are legitimately rare and should offer as much as a 3 to 1 leverage in the unfolding bull market in gold. Here's a brief run down of the coin types that make up this 8 piece U.S. gold type set and the rarity and price you can expect to pay at this point in time.

Coin #1: Gold $2½ Liberty (1840-1907)

The $2½ Liberty Gold coin was one of the backbones of America's monetary system for 67 years. In many parts of the United States, paper money was NEVER accepted in payment of goods and services. Merchants commonly insisted on payment in gold or silver. The Liberty series was designed by Christian Gobrecht and actually outlived the famous sculptor by 60 years!

This long, long series is filled with terrific rarities, but there are numerous late dates that survived in remarkable condition. A nice supply of post-1900 Liberty $2.50 gold coins survive in Gem Uncirculated MS-66 examples which have been graded by NGC or PCGS and now sell for about $2,200 each. These $2.50 Liberty gold coins were struck at 5 different U.S. Mints during the life of the series: Philadelphia (no mint mark), New Orleans (O located on the base of the reverse under the eagle), San Francisco (S located on the base of the reverse under the eagle), Charlotte (C located on the base of the reverse under the eagle) and Dahlonega (D located on the base of the reverse under the eagle).

Rare dates will give you added upside, and when completing a Gem Uncirculated U.S. Gold Type Set with better date coins, you stand the chance of obtaining an extra premium for the entire set.

NGC and PCGS have graded a total of 3386 $2.50 Liberty Gold Coins as MS-66. Higher quality coins are available: 405 have been graded MS-67 and 25 MS-68. MS-69s are extremely rare—they trade for $75,000 when available. MS-70 coins are non-existent.

Coin #2: $2½ Indian (1908-1929)

The $2½ and $5 Indian gold pieces introduced the wonderfully beautiful incuse design to American numismatics. The word "incuse" is so obscure outside of numismatics that it isn't in some dictionaries, but in coin terminology it means that the devices of the design are recessed rather than raised in relation to the fields. Dr. William Sturgis Bigelow, a close personal friend of President Theodore Roosevelt, got the idea for these incuse designed coins from some of the coins minted by the Egyptian Fourth Dynasty.

Bigelow obtained the go-ahead from Roosevelt to persuade the famous Boston sculptor Bela Lyon Pratt to submit models in this technique. Pratt's Native American chieftain model remains unnamed, his tribe unknown. Roosevelt enthusiastically approved the designs and ordered Pratt's models sent to Philadelphia Mint for translation into master dies. The coins were minted predominantly in Philadelphia but coins were also minted in Denver (D located left of the base on the reverse under the eagle) in 1911, 1914, and 1925.

The rarest coins in this $2.50 Indian gold series include the 1911-D, 1914, 1914-D. Issues in MS-66 are VERY rare and expensive. The 1909, 1910, 1912, 1913, 1925-D, 1929 dates are also rare and—we believe—are VERY undervalued in Gem Uncirculated MS-66 condition. NGC and PCGS have graded just 351 of these beautiful coins MS-66. The 1908, 1925, 1926, 1927, and 1928 examples of these coins are still rare but can be bought in MS-66 condition for about $7,500 at the present time. MS-67 and better coins are GREAT rarities, seldom offered on the rare coin market.

Coin #3: $5 Liberty, Motto (1866-1908)

The Congressional Act of March 3, 1865, which authorized the coinage of shield nickels and issue of certain classes of interest bearing notes, also ordered that henceforth all U.S. Coins large enough to provide room must include the motto IN GOD WE TRUST. The U.S. Mint took this to mean the half eagle, eagle, and double eagle, i.e. $5, $10 and $20 U.S. gold coins. A "no motto" version of the $5.00 Liberty gold coins was struck from 1839 through 1866. But for the purpose of discussion, we will reference the more common but still rare "motto" type.

On April 15th of 1933, just one month after his March 4th inauguration, President Franklin Roosevelt signed an executive order that made gold ownership illegal and compelling everyone to surrender "all gold coin, gold bullion and gold certificates" by May 1st, 1933. Exempt were true numismatic coins at the time. The Liberty $5.00 were NOT considered numismatic, so during the gold recall of 1933, huge quantities of $5 Liberty gold coins were shipped to Europe by hoarders in order to evade this order. In the 1970's many of these coins made their way back to the United States and were purchased by collectors. Swiss banks sold as many as 22,000 pieces in one gigantic sale. Not surprisingly, few coins survive in Gem Uncirculated condition.

NGC or PCGS Gem Uncirculated MS-66 examples now sell for about $5,000. These $5.00 Liberty gold coins were struck at 5 different U.S. Mints during the life of the series: Philadelphia (no mint mark), New Orleans (O located on the base of the reverse under the eagle), San Francisco (S located on the base of the reverse under the eagle), Charlotte (C located on the base of the reverse under the eagle) and Dahlonega (D located on the base of the reverse under the eagle). The coins struck from 1900 through 1908 were struck at just the Philadelphia, San Francisco, and Denver mints.

There's one neat variety coin we love: the 1901-S/O. The S mint mark actually stands over an O mint marked on the reverse of the coin under the eagle. There are just a few MS-66 examples of this coin. Only 1 graded by NGC and 1 graded by PCGS as of this writing. They're probably worth $9,500 -$10,500. Of the balance of the coins in this 1900-1908 period, NGC and PCGS have graded just 556 coins MS-66!

These coins can be acquired in MS-67 condition, but are EXTREMELY rare in MS-68 and MS-69 condition.

Coin #4: $5 Indian (1908-29)

The $5.00 Indian series has the same genesis as the $2.50 gold coins discussed earlier in this chapter. However this series is MUCH rarer, especially in MS-66 condition. The incuse design made it very difficult for Choice or Gem Coins to survive. These coins are most often available from rare coin dealers and at auction in Brilliant and Select Uncirculated condition (MS-60 through MS-63 grades), and still trade at substantial prices.

NGC and PCGS have graded less than 1000 coins in MS-65 condition and just 75 coins in MS-66 condition. There are many rare dates and they can trade for extremely high prices. A 1915-S in MS-65 recently traded hands for over $70,000. A 1911-D traded hands for $125,000!

Consider the 1908, 1909, 1909-D, 1911, 1912, 1913, 1914, 1915 in MS-65 or MS-66 condition. MS-67 and better coins are GREAT rarities, seldom offered on the rare coin market.

These coins were struck at the Philadelphia, Denver, and San Francisco mints and only struck one year at the New Orleans mint, in 1909. This is the great rarity of the series and worth over $200,000 in MS-66 and $350,000 in MS-67 condition!

Coin #5: $10 Liberty, Motto (1866-1907)

The Liberty $10 gold coin has the same origins as its $5.00 cousins, with one big difference. These coins are much harder to find in Gem condition. The greater surface area of this larger coin was more susceptible to scratches and abrasions. So, there are far fewer coins grading MS-66 by either PCGS or NGC. Also adding to this problem is the higher denomination. Again, who had $10 to put away and save? Ten dollars was a great deal of money back at the turn of the century. Collectors who had money tended to hoard lower denomination coins like $2.50 gold, which explains why NGC and PCGS has graded 3,386 $2.50 coins in MS-66 and only 480 $10.00 gold coins.

Fortunately there have been a few small hoards of these beautiful coins uncovered over the past 30 years, approximately 1000 that grade MS-65. Nearly 700 of those coins are dated 1901 or 1901-S. If it weren't for these hoarded dates, you'd be paying about $12,500 for MS-65 examples instead of $3,250. Gem quality MS-66 pieces are very scarce and sell for about $5,750 when available. We recommend MS-66 coins, but also like the upside potential of Choice Uncirculated MS-65 coins. These coins were also struck in Denver, San Francisco, and Philadelphia from 1900 through 1907. MS-67 grade coins are EXTREMELY RARE. As for higher quality coins, they are priced far beyond sensible collection prices for the average investor.

Coin #6: $10 Indian (1907-1933) "Eagle"

These $10 Indians come in two types: with motto and without. For the purposes of an 8 piece gold type set, either type would fit the bill.

There is one "hoard" date. The 1932 issue makes up more than 50% of the coins that survive in Choice Uncirculated MS-65 condition. Yet even the 1932 is really rare in Gem Uncirculated.

The $10 Indian was designed by one of America's most famous sculptors, Augustus Saint-Gaudens, and was designed along with the $2.50, $5.00 and $20.00 gold designs requested by Teddy Roosevelt. The Native American Indian woman bears of all things a feathered war bonnet. Historically, no Indian woman would have ever worn such a headdress. The Eagle that adorns the reverse of these coins is reminiscent of the breathtaking sculptures of eagles that decorated both early Egyptian and Roman coins and buildings. The edge of the coins minted from 1907-1911 features 46 stars, representing the states in the union at the time of its design. In 1912, two more stars were added, in honor of two additional states, New Mexico and Arizona.

NGC and PCGS have graded a combined 3,439 "Motto" and "No-Motto" $10 Indians in Choice Uncirculated Mint State MS-65 and only

610 in Gem Uncirculated MS-66 condition. These coins are VERY rare in Gem MS-67 and MS-68 condition and are virtually impossible to locate in MS-69. Nevertheless, I don't recommend buying the 1932 dated coins in anything less than MS-66 condition. The next most common coin is the 1926, which is a superb coin to buy in MS-66 if you can. There are many key dates in this series. These $10 Indians were struck at just three mints: Philadelphia, Denver, and San Francisco.

Coin #7: $20 Liberty (1849-1907) "Double Eagle"

In 1849 Congress authorized the minting of the largest gold coins the United States would have for regular circulation. These wonderful gold coins were struck at 5 U.S. mints: Philadelphia, Denver, New Orleans, San Francisco, and Carson City, Nevada.

The $20 Liberty was a symbol of wealth around the world and was in consistent demand around the world as a unit of exchange. Because of its substantial value, few of these coins remain in Choice or Gem Uncirculated states of preservation. For an 8-piece type set, let me focus on the coins struck from 1900-1907. Only 4,951 coins have been graded MS-65 by NGC, and PCGS has only graded 1,991. Only 270 have been graded MS-66.

So, it's with little hyperbole that we stress this coin will be the hardest to obtain in either Choice or Gem Uncirculated condition. The most common surviving coins from this series, post-1900, are the 1904 Philadelphia. Stay away from these 1904 common dates in Choice MS-65 condition, and pay the premium for a 1900, 1901, 1902, 1903, 1906, or 1907. Paying an extra 20-30% in this case should pay off nicely as gold takes off and sets the rare coin market on fire. When locating an MS-66, a 1904 is just fine.

Coin #8: $20 "Saint Gaudens" (1907-1933) "Double Eagle"

The "Saint Gaudens" $20 gold coin is the most famous gold coin in history. As of this writing the 1933 example of this magnificent coin holds the world's record high price at auction, selling for $7.5 million!

After FDR's executive order making private gold ownership illegal, all of the 1933 $20 U.S. Gold coins, which were still supposed to be at the mint (as none were released), were believed to have been melted by the mint and the gold that was yielded transferred to Ft. Knox.

However, it has been speculated and rumored for the past six decades that one or more examples were smuggled out of the U.S. Mint before they were melted. Some numismatic historians believe this coin was not smuggled but was in fact given as a gift by mint officials to King Farouk, Egypt's last monarch. Farouk is one of the last century's most famous rare coin collectors.

This coin re-emerged in 1996 and came into the possession of Stephen Fenton, who is a well known and respected rare coin dealer based in London, England.

Fenton smuggled the coin into the United States and was apprehended with the coin after attempting to sell it to undercover Secret Service agents in New York. Since the origin and the legality of this 1933 Double Eagle was clearly clouded and in question, the U.S. Government agreed to an out-of-court settlement with Fenton last year, the terms of which mandated the sale of the coin at auction and proceeds to be split. Fenton also agreed to pay an additional $20 cash back to the U.S. government in order to balance the books.

The sale price realized was $6.6 million for the U.S. government for the 1933 Double Eagle, a 15 percent commission for Sotheby's, and the coin's $20 face value.

This sale was an important turning point for rare coin collectors and investors. It showed that serious money is making its way from Wall Street back into hard assets and collectables.

We don't recommend the 1933 Saint Gaudens $20 or any others that might re-surface (rumors persist of 2-4 additional coins). Instead, consider the Gem Uncirculated MS-66 or even a Gem MS-67 example, dated from 1907 to 1932. There have been many hoards of these coins sold into the market. These $20 Gold coins were very popular in Europe and when FDR signed his executive order, many of these coins found their way to European banks. They were the most common balance of payment accepted by European banks for U.S. debt.

A final word for these wonderful "Saint Gaudens" $20 gold coins. In 1907 a small number of these coins were struck in High Relief. These coins are magnificent. If you start buying gold coins, you'll want one of these. Each coin was struck several times so Miss Liberty literally pops off the coin, much like the coin designs of ancient Rome, Egypt, and Greece. These coins trade for as much as $27,500 in Choice Uncirculated and $45,000 in Gem MS-66 condition. They come with a "flat rim" and "wire rim." PCGS and NGC together have graded just 146 coins of these two types.

Proof Coins

If you do buy an 8-piece 20th Century U.S. Gold Type Set in Gem condition, you'll run across the term "Proof."

A proof coin is made from specially selected coin blanks that have been highly polished and dies that are also highly polished. The coins are hand-fed in a slow moving press at the various U.S. Mints. They are struck multiple times to make sure the coins' details are superior to those that are made for circulation.

The Liberty gold coins in proof condition have high mirror surfaces. The Indian gold coins were produced with matte surfaces. The 20th Century gold coins discussed in this 8 piece type set are VERY rare in Gem Proof condition. Do not consider them as an investment unless you are investing $250,000 or more in the gold rare coin market. They have tremendous investment potential, but should only be sought after you've completed the 8-piece 20th Century Type Gold Set.

Conclusion

This book strongly encourages investing in gold to hedge yourself against the potential financial crises the U.S. and global economies face. It is best to invest directly into gold because of bartering capabilities, ease of transportation, available weights, and costs. Rare gold coins provide the best leverage.

However, for investors interested in leveraged and diversified gold investments, we'll now explore other venues for investing in gold.

CHAPTER 12

Indirect Investments in Gold: Gold Stocks and Mutual Funds

Some investors prefer investing in marketable securities because of their liquidity, low transaction costs, and industry regulations. We do strongly encourage you to always have some physical gold in your portfolio and rare gold coins are probably the best way to do this. Gold mutual funds could be a suitable alternative for some investors, such as IRA investors or investors on a budget. This section is not going to be a complete guide to mutual fund investing, but we will try to give you some sound advice that you can use to help you select a gold mutual fund that meets your needs.

A mutual fund is a fund operated by an investment company that raises money from shareholders and invests in actual gold, stocks, bonds, options, or money market securities and investments that are authorized by the fund charter. A professional money manager actively manages the fund. The investor has a share of the professionally managed portfolio. Mutual funds issue and redeem shares on a continual basis. They can be purchased directly through the mutual fund company or through a broker. Most are not available on an exchange, although exchange traded funds (ETFs) are becoming more prevalent.

Many investors have a significant amount of their investable funds in retirement accounts. Retirement accounts can't invest directly into gold because of investment regulations, so gold mutual funds are an alternative. The best alternative is the gold exchange traded trust that we will discuss later in this chapter.

How Do You Make Money in Mutual Funds?

Mutual funds work differently than owning gold or a stock. Investors in mutual funds can make money in three possible ways:

1. Price appreciation—An investor would buy shares in a fund. The portfolio could consist of actual gold, gold mining stocks, diversified mining stocks, and cash. The share price would be the value of the portfolio minus management and operating

expenses, divided by its outstanding shares. In this example, we could say that at the beginning of the year the portfolio value minus expenses was $100 million, divided by their 10 million shares, giving a value of $10 a share. This $10 is considered the net asset value, NAV. The NAV is quoted in most newspapers. Gold continues to rise and the value of the shares rise to $11.00 by the end of the year. The appreciation of the fund would be 10%.

2. Capital gains—You have to remember that your mutual fund is actively managed. Let's say that the portfolio manager sold some stocks in the portfolio, some at a loss and some for a gain, for a net capital gain of 50 cents per share. This would give you a capital gain of 5%.
3. Dividends—Some of the stocks and money market securities may throw off some income. In our example let's say it's 17 cents.

Here is an example of the break down of the potential returns in a mutual fund:

Appreciation $10 to $11 = 10%
Capital Gain Distribution of 50 cents = 5 %
Dividend and income of 17 cents = 1.7%
Total return is 16.7%

A prospectus will break down the total return for you. Some online resources will also detail the total return.

SEC regulations require that the dividends and accumulated capital gains will be "distributed" to shareholders periodically. Shareholders have a choice of taking the distributions in cash or in the form of additional shares in the fund. Check with your fund to learn about the distribution choices your fund offers. Be aware that the gains will have been accruing in the fund along the way, but once the distributions are made the fund will have a lower NAV, similar to a stock that goes ex-dividend.

In the example above the NAV would drop by 6.7% (capital gain and dividend distribution). A person that decides to take a cash distribution will have a lower NAV, a person that reinvests will have more shares and a lower NAV. It is the portfolio manager's goal to increase the NAV. Mutual funds work ideally in retirement accounts. If you are investing in mutual funds in a taxable account, the capital gains reporting can get a little tricky—so make sure you consult your tax advisor or financial planner.

Benefits of Gold Mutual Funds

Mutual funds are a very good vehicle for the small investor, IRAs, and other retirement accounts. It would be very difficult for a small investor to duplicate the diversification and professional money management of a mutual fund. We have listed several benefits of mutual funds below:

1. Diversification—Mutual funds give the investor the ability to own a share of a much larger portfolio that is also diversified. This could be a plus and a minus. The main idea behind diversification is to reduce your risk by not having all your eggs in one basket. But there are times where money managers diversify too much and the benefit of gold is diversified away. They may invest in treasuries, companies that provide services to gold mining companies, and companies that mine not only gold but other precious metals like silver and platinum. All of this may be good or bad, depending on conditions.
2. Professional Money Management—To be a truly good stock picker takes education, intelligence, experience, training, and talent. You need to understand business, economics, accounting, finance, marketing, leadership, management, and the dynamics of the market. To be a good gold stock picker, you'll also need experience and knowledge of the gold industry, and it also helps to have gold industry contacts and relationships. When choosing a mutual fund, make sure its money manager has the right qualifications.
3. Most mutual funds have low investment minima, some as low as $50 for IRA accounts. This can be important for young investors or investors on a budget.
4. Services—Many mutual fund companies provide shareholder services. Services include quarterly statements, dividend and reinvestment plans, semiannual updates from the fund manager, customer service support via 800 numbers, and automatic investing plans. Make sure you are familiar with all the services your mutual fund offers by asking your broker, or calling the mutual fund company.

There are a few drawbacks to mutual funds. Funds do not always perform like gold because they have securities and other non-gold investments. We pointed out that there are costs and fees involved with gold funds. You have to be very careful and understand all your costs as they can be a drag on performance.

One concern that investors have had to worry about lately is the consolidation of the mutual fund industry and fund manager changes. You may buy a fund because of the performance of the money manager and that manager could leave; you are then left to decide to sell, or wait and see how the new manager performs. There has been a lot of consolidation among mutual fund companies, so the company, fees, services, and portfolio manager could change in the future.

How to Select a Mutual Fund

One of the advantages of investing in this gold bull market vs. the last one is that there are a lot more mutual funds available now. But this is also one of the disadvantages: there are too many funds to choose from. Fortunately there are online services that can help you narrow your choices so that you can select a fund that meets your needs. Here are some important criteria to consider:

Performance—This is obviously the most important factor to consider. Many services will give you a 10, 5, 3 year track record. The 10-year track record is the most important to review, as it will include good and bad years and at least two business cycles (most business cycles last 3-5 years). You also want to look at years where gold performed poorly such as 1996-1998. How did the managers do in a bad period compared to the actual price of gold and other fund managers? Most money managers can make money when gold is moving up, but the really good managers distinguish themselves in bad gold markets by protecting the portfolio. There are many defensive measures a fund manager can take to protect the portfolio: move to cash or treasuries, invest in gold stocks that are diversified, and hedge with options or futures (not all funds allow their managers to hedge).

Try to make sure that the fund you are interested in has at least a 5 year track record. That way you can judge how they perform in good and bad years.

When comparing performance of these funds, you should ignore ratings that compare gold funds to the S & P 500 or some other broad index. Gold funds are a specialty asset and are often uncorrelated to the stock market (which is what you want). Compare them instead to other gold funds.

Management—Normally if the fund has good long-term performance, it will have a good manager. The qualities of a good gold money manager include experience in the gold industry, intelligence, education, training, investment experience, contacts in the industry, and knowledge of the markets. Many mutual fund companies consider an MBA from a top tier university and a CFA (Chartered Financial Analyst) designation as a minimum education requirement for their money managers. This is a very good idea since this background gives them a solid foundation in evaluating businesses, the economy, and corporate valuations. It's also important that the gold money manager has either worked for a gold mining company or has an undergraduate degree in geology or engineering. You can learn about the manager's background in the prospectus, or from Morningstar mutual fund reports.

Also compare the fund manager's tenure and the fund's performance. If it does have good performance, is the fund manager who guided the fund to the good performance still the fund manager?

Diversification—With most mutual funds, diversification is a good thing. It allows you to participate in different sectors in the economy as well as reducing risk. For gold mutual funds, too much diversification can make the fund act more like a financial asset and less like gold. There are online tools that can help you determine the composition of the fund. Make sure that there are not too many stocks and that the companies it invests in are involved directly with gold. Also make sure that there are not a lot of diversified mining companies or suppliers to mining companies. We have provided a list of mutual fund companies at the end of this chapter, and some will have "precious metals" as part of their names. These precious metal companies may be funds you want to avoid. There are some merits of a diversified precious metals fund as the flexibility could work in the long run, but to take advantage of the forecast we see, you want a fund that mimics gold as much as possible.

Also make sure that the turnover—the number of times each year the portfolio changes—is not high, and there are not high levels of cash. It does make sense for the money manager to change the portfolio to be more defensive when gold is in a bear market, but not in this environment.

Fees and loads—Fees can be broken into two types: management fees and loads (commissions). All funds will have management fees, but not all funds will have a load.

Management fees can be less than .5% or higher than 2% per year. The fund calculates its returns and fund value net of fees; you are not charged fees directly. A high fee structure could certainly be a drag on performance. There are very few fund managers whose performance is good enough to overcome high fees.

When it comes to loads, there are basically 3 ways to buy funds:

1. No load: Many financial publications advise no-load funds as they allow you to buy the fund at NAV without commissions. If you're willing to do your own research and you have the experience and knowledge to select a mutual fund, then we don't have to tell you that a no-load is probably your best choice. If you go directly to the fund, its staff can help you learn more, but they will probably

not be objective about their own fund. Some discount brokerage firms do have staff with expertise to help you select a fund that may meet your needs. If you get some help from a discounter, make sure you ask someone with the appropriate experience and background. There is plenty of on-line help available for learning about mutual funds and how to select a fund that meets your needs.

2. Load: A load fund is sold at the NAV plus commission, but redeemed at the NAV. Let's say there is a 5% load and the NAV is $10. You would add 5% to the NAV, so you would pay $10.50. Normally load funds are sold by brokers. Most brokers at the big firms do not offer no-load funds. If you have a broker or financial adviser that you trust and have a good relationship with, your broker can help you select a fund that would be appropriate for you. Make sure you hang on to the fund for a long period of time so you can spread the cost of the load over many years.
3. Back-end loads: To compete with load funds, brokers introduced back-end load funds. You buy the funds at NAV and they have a sliding scale charge to sell. They normally start at 5% for the first year and slide to no charge to sell if you hold for more than 5 years. There's one caveat with these funds: their management fees are normally much higher than no-load or load funds.

A subset of the back-end loads are funds that charge a redemption fee if the fund is sold early. These fees are normally a lot lower. A typical redemption fee would be 2% if the investor sells before 60 days, usually with no redemption fee afterwards.

Mutual Fund Resources

There are excellent resources and online tools to help you find and analyze mutual funds. The most well known resource is Morningstar. We remember when Morningstar started in the late 80s. They have come a long way, with great improvements in the resources they have for mutual fund investors. Their website has some free services, and there are some subscription services if you decide to become a heavy mutual fund investor. Their website is www.morningstar.com.

Yahoo has an excellent mutual fund service in their Yahoo finance section at http://finance.yahoo.com. MSN money also has a useful mutual fund section. Many online brokers and traditional brokers have online mutual fund screening tools and information. Most brokers and discounters have free literature to help investors learn about mutual funds. Ask your broker what they have available.

Don't forget to read the prospectus. Regulators have forced mutual fund companies to make their prospectuses investor friendly. There is a wealth of information—you can learn a lot about a fund, its historical performance, objectives, fees, background information of the portfolio manager, and risks.

Mutual Fund List

Below is a list of mutual funds that can help you select a fund that meets your needs. There are more funds available, but some were too small, performance was not competitive, or they were closed to new investors. You will notice that many of the big name mutual fund money managers (Fidelity, Vanguard, Franklin, Oppenheimer) have a gold or precious metals fund. The list focuses on performance with the top 5 year performers listed first.

Fund Name	5 YR	3 YR	10 YR	Best Year % (2002)	Worst Year % (1997)
Tocqueville Gold	28.11	46.79	N/A	83.3	-10.7
First Eagle Gold	25.76	52.99	N/A	107	-29.8
USAA Precious Metals & Minerals	25.5	51.52	7.29	67.6	-38.2
Gabelli Gold	24.29	47.21	N/A	87.20%	-51.9
Evergreen Precious Metals	23.05	47.68	3.92	72.4	-13.4
Vanguard Precious Metals	22.9	34.36	4.69	33.4	-38.9
Franklin Gold & Precious Metals	19.29	28.25	4.26	37.4	-35.7
Oppenheimer Gold & Special Minerals	18.52	34.15	5.83	42.3	-31.9
ING Precious Metals	18.35	40.23	3.05	66.7	-43
AXP Precious Metals	17.62	36.02	5.6	60.8	-49.3
Fidelity Select Gold	17.23	35.85	4.16	64.3	-39.4
American Century	17.2	45.2	1.31	73	-41.5
US Global Investors Gold Shares	16.15	45.01	-10	81.4	-57.4
Van Eck International Investors	16.11	44.97	0.49	91.3	-36

Invesco Gold & Precious Metals	14.99	36.15	-1.49	59.7	-55.5
Rydex Precious Metals	11.97	30.84	-1.65	48.2	-37.6
Midas	6.1	33.48	-5.18	61.1	-59

Figure 12.1: Gold Mutual Funds

Closed-End Funds

Another way to invest in gold is with closed-end funds. These funds have a fixed number of shares and sell on exchanges. They do not issue and redeem new shares everyday like a mutual fund, although they might make a secondary offering if demand is high. They tend to specialize in a sector, country, or asset. Closed-end funds trade at a discount or a premium to their NAV.

There is a closed-end gold fund called the Central Fund of Canada (ticker symbol CEF) that invests exclusively in gold and silver bullion. The fund's objective is to hold at least 90% of its assets in gold and silver physical bullion. At the time of this writing the fund held 297,000 ounces of gold and almost 15 million ounces of silver. But one of the problems is that the fund currently trades at approximately a 15% premium to the NAV, and this premium fluctuates.

As you can see below the fund does follow the price of gold fairly closely. The risk for the investor in CEF is not only that the price of gold could move adversely, but also that the premium above NAV could contract, magnifying potential losses.

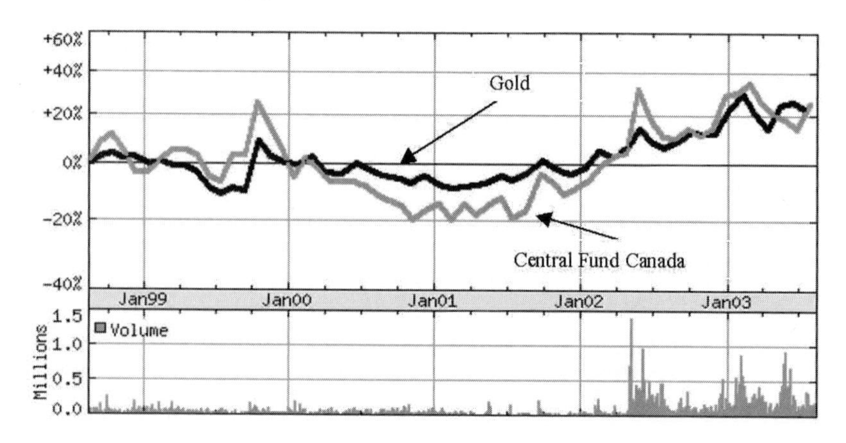

Figure 12.2: Central Fund of Canada Price Performance vs. Gold

The premium/discount of CEF's share price to NAV tends to expand and contract along with rallies and declines in the price of gold. Some savvy gold traders have learned to use the level of this premium as a contrarian sentiment indicator for changes in gold prices.

Equity Gold Trust Fund (EGTF)

This is an exciting new gold investment that will soon be available to investors. The EGTF will invest in actual gold bullion. This investment will be the next best alternative to owning gold. If you are familiar with QQQs, Diamonds and SPDRs, then this will be gold's equivalent.

EGTF will be very different from the traditional gold mutual fund. It will not be diversified, it will not invest in gold mining companies, and it will not be professionally managed. The trust would own gold bullion and its shares are expected to be priced at $1/10^{th}$ of an ounce of gold.

At the time of this writing, that would be approximately $40.00. If you bought 100 shares it would cost you $4,000 plus a brokerage commission. It is planned that the actual gold will be deposited in vaults of the Hong Kong Shaghai Banking Corp. The shares are expected to be traded in London and Tokyo stock exchanges, and potentially in Toronto and Johannesburg. In the U.S. they will trade on the New York Stock Exchange under the symbol GLD. The management fee is expected to be a very low .12%.

The advantage to the investor is that you don't have transportation, warehousing, or security costs for buying and storing your gold. The fund takes care of that for you for a very modest fee. You also do not have the typical markup that must be paid to a dealer for buying bullion.

At the time of this writing the EGTF shares are waiting for the Securities and Exchange Commission's approval. The World Gold Council expects this product to be available in the second half of 2004.

Institutions and retirement accounts held by individuals are barred from owning gold directly, and they control trillions of dollars of investment funds. But by making gold available in the form of a stock, the EGTF will be eligible for these accounts. The demand for gold will explode with this event alone.

If you're going to invest in gold funds the EGTF is probably the best way to invest.

Investing in Gold Stocks

The best way to invest in gold stocks is to seek help from an experienced, knowledgeable, and honest stockbroker. You can sometimes get gold stock tips from investment newsletters and some newsletters specialize in gold and gold stocks.

(One of the authors of this book, James DiGeorgia, has a gold investment advisory service. He is also one of the most experienced rare coin dealers in the country. We'll give you more details about his advisory service and his coin service at the end of this book.)

There is no way we can teach you to be a good gold-stock picker in one chapter, but we'll cover a few basics.

Investing is gold stocks is different than investing in traditional stocks. The main determinants of stock price movement for traditional stocks are interest rates, inflation, earnings, and the economy. Falling interest rates and inflation, a thriving economy, and growth in earnings are key to price appreciation for most stocks.

As we have discussed in the economic section, gold appreciates when there is inflation, instability and geopolitical threats, and also an imbalance in supply and demand. Gold stocks normally appreciate when gold increases in price. With gold stocks, the keys to focus on are the gold reserves of the company, along with the potential impact that the price of gold could have on earnings.

The traditional stock metrics like P/Es and growth are not as important with gold stocks. As a matter of fact, most gold stocks have very high P/E ratios and will look very expensive. With most stocks, growth in earnings are what most investors are looking for, and they look at P/Es to determine how much they're paying for those earnings. What investors forget about gold stocks is that many gold stocks have assets, and those are assets are gold. The gold reserves are an important variable in determining values of gold stocks.

Here are some of the variables that impact gold stock prices:

- The price of gold
- Inflation and interest rates
- The supply and demand for gold
- Gold reserves
- The cost to produce an ounce of gold
- Quality of management
- International vs. domestic—International stocks can be priced lower due to international risks which are discussed later in this chapter.
- Diversified vs. non-diversified—When gold is rising, non-diversified gold companies will normally do better than diversified companies. Diversified companies may have operations in other

metals, like silver and copper, and therefore will not move as much as gold stocks that focus exclusively on exploring and mining for gold.

Gold Reserves and the Cost of Gold

Probably the most important quality a gold company should have is high gold reserves. As the price of gold rises, the companies with the highest gold reserves could become the most valuable. Reserves are labeled as *proven, probable,* or *possible.*

Securities regulators and the accounting profession have rigorous rules that govern the labeling of reserves. To be *proven*, drilling and sampling of ore must be closely spaced. Since this is an expensive process, it is likely that mines will only prove a few years of production. Because of this, *proven reserves* are not a good indicator of the true life of the mine. *Probable reserves* are less rigorously defined. Remember, *proven*, *probable*, or *possible* are only estimates.

The company's cost to produce gold is another important criterion to consider. North America has a relatively low cost to produce gold. There are many factors that determine the cost of producing gold: labor, location, accessibility, the depth of the deposits, the hardness of the rock, the amount of waste material that must be removed to mine the gold, and the cost of compliance with environmental regulations.

You can start your search for gold stocks by using stock screeners that are available on most online brokerage accounts and free sites like Yahoo Financial. For example, we used an on-line broker stock screener and came up with 80 domestic and international gold stocks. The table below lists the biggest market capitalization companies from the list.

Company Name	Market Cap (Mill. $)	Cash Cost ($/oz)	P&P Reserves (Mill. oz)
Anglo American plc (ADR)	23271.49	203	72.30
Newmont Mining Corporation	11525.06	155	87.20
Barrick Gold Corp.	9858.98	177	86.90
Gold Fields Limited (ADR)	5867.43	183	79
Placer Dome Inc.	4893.7	195	52.90
Harmony Gold Mining Co.	2567.75	225	49
Kinross Gold Corporation	2190.31	220	13.20
Goldcorp, Inc.	2145.02	99	5.50

Compania de Minas Buenaventura (ADR)	1793.03	180	1.10
Glamis Gold Ltd.	1489.34	170	5.70
Meridian Gold Inc.	1118.54	87	4.20
IAMGOLD Corporation	737.65	133	7.10
Randgold Resources Ltd. (ADR)	490.22	74	11.50
Hecla Mining Company	446.31	137	7.69
Eldorado Gold Corporation	367.80	230	5.80

Figure 12.3: Gold Stocks
[P & P = Proven and Probable]

(There are also several caveats we must mention about the above list. First, some of the companies listed produce more than gold. Some also produce silver, platinum, copper, and other base or precious metals. Many of the companies are domiciled outside of the Unites States. International investing has currency, political, and regulatory risks that are normally higher than U.S. investing. Make sure you understand all the risks with international investing and the risks associated with the country you are investing in.)

From this table we can see the companies that have the highest reserves and lowest costs, and which companies the market is valuing the most. The companies that have the lowest costs have the best chance of being the most profitable. The companies with the largest reserves have the potential of being the most valuable, especially if the price of gold increases.

Let's try to understand what each column means. The first column is the market capitalization. Price is actually the market capitalization of the company: the current stock price multiplied by the company's shares outstanding.

For instance, if you wanted to buy Microsoft you would multiply the current price, $26.47, by its shares outstanding, 10.8 billion shares. To own Microsoft you would have to pay $285.8 billion, its market capitalization.

[Microsoft $26.47 (current price) x 10.8 billion shares outstanding = $285.8 billion market capitalization]

Now let's look at the stock on the list which has the highest market capitalization, Anglo American. To own Anglo American you'd have to pay about $23 billion.

[Anglo American $15.77 (current stock price) x 1,476,370,944 shares outstanding = $23.28 billion market capitalization]

The next column is the cash cost to produce an ounce of gold. Anglo American's cash cost is around $203. It's not the highest or lowest on the list. Rangold has the lowest cost, $70, so Rangold has the potential to be the most profitable of the group as the price of gold rises. This is easy to figure out. If the price of gold is $350 then Anglo American would make $147 per ounce of gold, or a 72% return. If Rangold sold an ounce of gold they would make $280 per ounce, which would be about a 400% return. Some investors would be interested in the potentially most profitable companies which have the lowest cost.

The next column is the companies' gold reserves. As we discussed earlier, these reserves are estimates only, and also it would take years to mine all the reserves.

Let's take a look at Newmont Mining, a U.S. company. They have the highest gold reserves, proven and probable, on the list. If the estimates are accurate and they could mine all the gold, the company would be worth approximately $30 billion ($350 gold price x 87.20 million proven/ probable gold reserves). The market capitalization of Newmont is $11.5 billion. In Chapter Eleven we discussed gold forecasts of up to $8000. Let's just assume, however, that the price of gold reaches $1,000. Then the potential value of Newmont based on its reserves would be close to $95 billion. At the current market capitalization, $11.5 billion, the cost to buy Newmont seems cheap using this valuation method and forecasts for gold.

The stock ideas above are just a starting place and it would certainly make sense to do your own due diligence on stocks that you're interested in.

There are also sophisticated and highly leveraged investments, like gold options. One associate of ours bought gold options at $.10 each that went to $5.00—making an astounding 50 times his money.

These, however, are for the very seasoned investor. For the average investor, or for anyone who simply wants to have a safety net on their financial future, we recommend direct investments or simple indirect investments like mutual funds.

And now, let's wrap all this up…

Part V

Safe Haven from the Gathering Storm

CHAPTER 13

The Rise of Gold in the 21st Century

From thousands of years ago in Egypt, where the use of gold was a royal prerogative, and goldsmithing became an art...

To the legend of the Golden Fleece, stolen by Jason and his Argonauts 1500 years before Christ...

To the kingdoms of Lydia, Persia, Athens, and Alexander the Great, where coinage began...

To Dionysius of Syracuse (405-367 B.C.), who restamped one-drachma coins to read two drachmas—and earned his place in history as one of the first politicians to debase a currency...

On down through the ages—through wars, conquests and the California Gold Rush, the largest migration of people in history, all in the quest for gold...

Right up to modern times, when mines go two and a half miles below ground, and central banks sell tons of gold...

For five thousand years, gold has been a universal currency and store of wealth. The only real question of its value has come in the last hundred years, with the ascendancy of paper currencies, vast credit balances, and the thinking in some parts that these systems made gold and gold-based currency standards obsolete.

We believe their thinking is unproven and highly questionable....

Where We Stand Today

Let's quickly review the economic dangers we face now:

- All the major world recessions of the last 30 years have had their roots in either disruptions in oil supplies or high oil prices. All of the great industrial economic powers of the world are very dependent on oil. The Middle East holds roughly 65% of the world's oil reserves, with almost 25% of the world's oil reserves in Saudi Arabia alone. The area is highly unstable, and vulnerable.
- We are much more dependent on imported oil than we were 30 years ago.

- The U.S. Federal Reserve Bank, "the Fed," is losing its power to manipulate the economy. Since 1995, it's pumped money into the economy to avoid economic problems caused by the Asian Contagion, the Russian debt default, the Long Term Capital Management crisis, Y2K, and to keep our economy going. Rates are now close to zero and potential bubbles are occurring in the bond and real estate markets. If we have another crisis, the Fed will not be able to lower rates to keep the economy going as they have done in the past.
- Derivatives still exist, are still being used, and could cause huge problems once again. Even Warren Buffett warned in his 2002 annual report that our economy faces grave potential dangers from derivatives. Buffett pointed out that many derivatives are carried on companies' books with overstated values. Part of the Enron mess was caused by their use of derivatives and overstating the values of these instruments on their balance sheets. Derivatives can also post enormous problems from their leverage and (lack of) liquidity.
- The U.S. budget deficits are much larger than the government acknowledges. Interest rates will have to be raised to attract foreign dollars to finance these debts. The weight of the debt and its interest payments will hinder the real and long-term potential of the economy. In the worst case the government could continue to print money to pay these debts and this could lead to much higher inflation and a much lower standard of living. Right now debt as a percentage of GDP is greater than the ratio during the Depression. And some experts say that by 2008, the future deficit could reach $54 *trillion.*
- Pumping huge amounts of money into an economy brings the problem of too many dollars chasing too few goods, i.e. inflation. If inflation and interest rates rise it could cause a collapse in bond prices. Also, with all this cheap money that's been available, speculators have been borrowing short-term to speculate in the markets. This can amplify market movements if rates do reverse. Once rates start to rise, borrowing costs start to go up, and asset prices start to fall. The speculators are forced to sell, creating even more selling in the bond and stock markets.
- The falling dollar has eroded international confidence in the dollar. A lower dollar is also inflationary. If the dollar gets too low, the Fed will be forced to raise rates to attract foreign money to finance our twin deficits: bad news for bond and stock investors.

There are now no major or significant currencies tied to gold. Currency values are determined arbitrarily by inflation, political stability, economic stability, investment opportunities in a country, national resources, and by government decree. But perhaps the biggest factors in currency valuation are inflation and interest rates.

Valuations can also depend on whether a country is an importer or exporter of goods. Currency valuations of countries that are importers are more vulnerable. The U.S. only imports about 12% of its needs—mostly oil. But that is a critical import—the one most likely to have serious problems.

Some of our biggest problems are:

- **The U.S. has been issuing huge amounts of debt...there's now $7 trillion in bonds outstanding...**
- **Inflation is inevitable...**
- **The stock market is still historically overvalued, at a P/E of 20; the historical average is 14 to 16...**
- **The U.S. still has too much production capacity that came online in the 1990s.**

But the wild card is oil.

Oil producers are getting less money for their oil, since they get paid in dollars and the dollar has been falling. And they're not happy about it.

Also, interest rates have been held artificially low. Eventually they're going to have to rise. The budget deficit is too big—over $500 billion.

Two more big problems are lower taxes and higher spending. How will the higher spending be financed?

Also, the Japanese have been buying U.S. treasuries, and now the Chinese have, too. But will they keep on, with the dollar falling?

A Feeble Fed?

Since 1987 Greenspan has been lowering rates to meet crises. But the Fed has run out of room to lower rates to offset problems. So what now? What will the government do in the next disaster? Look at Japan. As their economy tumbled, they lowered rates to the bottom. Then there was nothing to do but wait...for a long time...

At the time of the 1987 stock market crash, the discount rate was around 7%. Now they're at 1%. There just isn't much wiggle room left.

So what could happen that could spike gold prices way, way up? The answer is—any major disaster! A terrorist attack...a crisis of confidence in one of the major currencies...a major war...and the list goes on.

And then there's the X-factor...

China and India: The X Factor

There are over 2.25 billion people in India and China. People who are rapidly becoming more affluent and buying gold, and who historically have faith in gold.

That's a big market.

If even tiny portions of the new wealth being created in those countries go into the acquisition and holding of gold for safety, we're talking about a huge amount of gold purchasing.

Let's say even 1 out of every 10 of those people decided to keep some of their new-found wealth in gold.

That's 225 million people. Even if they only bought one ounce each, it's over 6,000 kilos of gold.

Some $90,000,000,000 ($90 billion) worth.

Bottom line, there are more people interested in gold, more ways to buy it than ever before, and more convenient ways than just holding bullion. With the advent of the EGTF (Equity Gold Trust Fund) and the ability to own bullion safely, there will be wider and wider acceptance of investing in gold. No more will investors have to worry about transportation, storage, security costs, dealer reliability, or other concerns.

Let's look one more time at the long-term chart of gold prices:

Figure 13.1: Long-Term Gold Prices

Going back 30 years, we see it's been almost 25 years since the last high. Since then, gold found support at the $250 to $300 area. But gold broke decisively out of its bear market in 2003, and has held above the breakout line.

It's also worthwhile to note that the bottom coincided with the peak in stocks, and gold rallied while stocks dropped.

And what happens now, if the stock market doesn't go up, but goes much further down?

In any event, we believe we're in for a good, solid, long-term bull market in gold. And if there's a disaster or a large-scale crisis in confidence in the dollar or other paper currencies, gold may be the only thing that saves your financial health.

All of which brings us to one last question: How much money to put in gold?

How Much Gold to Buy

The answer, of course, is that it depends on how much money you have and how much you desire a margin of safety.

The likelihood of a big problem sometime in the near future is high. It could be an oil shock, it could be a terrorist attack, it could be a crisis in confidence in one of the world's major economies, it could be natural disaster, it could be a major bear market.

> What you need is *protection.* Looking at it from a basic financial standpoint, 5% of your assets is simply not enough to provide you much protection. We could say that 10% is barely enough. And we could say that enough to make a difference would probably be 20% to 25% of your assets.

We're certainly not saying to run out and put all your money in gold or other precious metals. That would be extreme, and foolish at best.

But we are saying that it's prudent to have a sizable portion of your assets in gold.

For perspective, look at the recent historical snapshot:

In recent years, the Dow went from a high of around 11,700 to around 8,200—a slide of close to 30%.

But the Dow represents a small fraction of the stock market. It is a proxy for the market and the economy. A more realistic view is gained from the S&P 500—500 of the largest companies in America.

During that same period the S&P went from 1550 to 770—a 50% slide.

So what we're looking for is something that could hedge you for at least a portion of losses like that. And it's even more important if we face a disaster of greater proportions.

For a quick example of how gold can protect you, let's say the S&P drops 20%. If you have a $100,000 portfolio, it falls to $80,000.

But let's say you have 20% of your assets in gold ($20,000), and gold goes up 50%. First of all, because you've reduced your exposure to stocks by 20%, instead of losing $20,000 from $100,000 you now lose only $16,000. Secondly, you make $10,000 on your gold, so you've reduced your losses by over half.

That's on pure gold. If you're holding leveraged gold, you'll need less. And in a really big financial crisis, the differences could be much, much more.

Conclusion

Our first duty is to ourselves and our families. Just as we act prudently when we secure our houses for the night, drive carefully as we conduct our affairs during the day, and monitor our health and the health of our loved ones, we must also act prudently in our financial lives.

This means not only wise choices for today, but wise choices with ten and twenty year horizons.

We, the authors, wrote this book to repeat once again what others have said in different ways. Beware of the future. Protect yourself and your family.

There is one single best way to protect yourself. A way that will likely turn $100,000 into $200,000 or more in the next few years...and maybe much, much more. And that is the simple purchase of gold bullion. In 5,000 years, nothing of significance has changed the universal value of gold. It is revered today just as it was in ancient times.

Take a portion of your assets and purchase some gold. Better yet, do the two simplest things you can do to protect yourself and your family in the coming years. Pay your debt down and buy gold.

It could be your saving grace if just a few of the things we have outlined come to pass.

Spread your gold investments into a few other vehicles that give greater leverage, and this safety valve could even make you a fortune.

Buy gold now. Buy it over the next few years. Especially, buy on dips in the price.

It is the only investment that has proven itself over thousands of years as a store of value and a safe haven. And the chances of needing it are far greater than the chances of not needing it.

In fact, it's quite possible that gold will be "the investment of the century!"

Free Offer!

As purchaser of this book, you're entitled to a FREE gift: an introductory subscription to the Gold & Energy Advisor (GEA), a monthly newsletter covering the precious metals, diamond, and energy markets. As editor, James DiGeorgia delivers insightful commentary and investment recommendations, with the goal of delivering profitable recommendations to subscribers.

The GEA provides news, commentary, and recommendations that you won't get elsewhere. Large market-moving forces are developing that will dramatically affect your investments—but you won't hear about them from the mainline media. The GEA is dedicated to fully informing you about these forces and trends. (See http://www.goldandenergyadvisor.com for more info.)

The GEA normally costs $189 per year, but an introductory subscription is yours FREE! Call 1-800-819-8693 and ask for your free subscription.

About the Authors

James DiGeorgia is the publisher of several international financial newsletters (including the *21st Century Investor*).

James is an avid coin collector, and started buying and selling coins while still in high school. Over the years he's bought and sold well over $100 million in rare coins.

After earning a B.A. in economics, James worked as a numismatist for some of the largest precious metals and numismatic companies in the world.

In 1991 James accepted the editorship of the world-renowned and fiercely independent *Silver & Gold Report,* a precious metals advisory service. He became known as a tireless advocate of individual investors. James authored *The Insider's Guide to Buying Gold, Silver and Rare Coins,* and was frequently quoted as an expert in the *New York Times, USA Today, Los Angeles Times, Money* magazine, the *Chicago Tribune,* and *Barron's,* to name just a few.

Since then, James has founded a family of successful financial newsletters; see its website at www.21stcenturyinvestor.com.

Don Mahoney is an entrepreneur and investor who started investing in precious metals and watching the precious metals markets in the 1980's. He's a self-made millionaire who also worked for 11 years as a financial copywriter for various investment newsletters and services.

Don made hundreds of thousands of dollars in real estate, and owns property in New York, South Florida, and Central America. In recent years he has made money on platinum, palladium, and gold, and is currently building a personal collection of gold coins.

References

Introduction

http://encarta.msn.com/encyclopedia_761570498/Gold.html
http://www.geocities.com/mrgoldnugget/

Chapter One

http://www.bulfinch.org/fables/bull17.html
http://www.fordham.edu/halsall/ancient/asbook.html
http://www.fordham.edu/halsall/ancient/neo-babylonian-legaldecisions.html
http://www.greece.org/poseidon/work/argonautika/jason3_2.html
http://www.hyperdictionary.com/dictionary/talent

Bernstein, Peter L. *The Power of Gold: The History of an Obsession.* John Wiley & Sons, New York. 2000.

Buranelli, Vincent. *Gold: An Illustrated History.* Hammond Incorporated, New Jersey. 1979.

Vicker, Ray. *The Realms of Gold.* Charles Scribner's Sons. New York 1975.

Chapter Two

See endnotes below.

Chapter Three

Bradshaw, Greg, *Nazi Gold: The Merkers Mine Treasure,* Quarterly of the National Archives and Records Administration, Volume 31, No. 2 (Spring 1999)

Isaacs, Jeremy (producer), The World At War, Thames Television 1975

Johnson, Chalmers, *The Looting of Asia,* London Review of Books, 20 November 2003

Keegan, John, The Second World War

Koning, Hans, *Germania Irredenta,* Atlantic Monthly, July 1996, Volume 278, No. 1

Plaut, James S., *Hitler's Capital,* Atlantic Monthly, October 1946, Volume 178, No. 4

Plaut, James S., *Loot for the Master Race*, Atlantic Monthly, September 1946, Volume 178, No. 9

Rothfeld, Anne, *Nazi Looted Art: The Holocaust Records Preservation Project*, Quarterly of the National Archives and Records Administration, Volume 34, No. 3 (Fall, 2002)

Shirer, William L., The Rise and Fall of the Third Reich

Telegram to Secretary of State from "Gray, Berlin" dated November 28, 1941.

U.S. and Allied Efforts to Recover and Restore Gold and Other Assets Stolen or Hidden by Germany during World War II: Preliminary Study, May 1997— United States Department of State

Wise, Michael Z., *Reparations,* Atlantic Monthly, October 1993, Volume 272, No. 4

Chapter Four

http://216.239.57.104/search?q=cache:WFvuhlPF608J: www.econ.cam.ac.uk/faculty/meissner/gold_oeeh.pdf+Britain+%22 Gold+standard%22+history&hl=en&ie=UTF-8

http://econ161.berkeley.edu/TCEH/Slouch_Restoring11.html

http://en.wikipedia.org/Pound_Sterling

http://en.wikipedia.org/wiki/Boer_War

http://nsccux.sccd.ctc.edu/~mnutti/seas/presentation2.html

http://www2.sjsu.edu/faculty/watkins/hyper.htm

http://www.fff.org/freedom/0700f.asp

http://www.gold-eagle.com/editorials_02/wallybently100802.html

http://www.militarymuseum.org/Gold.html

http://www.milliondollarbabies.com/Frames1.htm

http://www.milliondollarbabies.com/Frames2.htm

http://www.milliondollarbabies.com/Frames3.htm

http://www.mises.org/money/4s3.asp

http://www.mises.org/money/4s5.asp

http://www.mwsc.edu/eflj/german/gc/inflation.html

http://www.traveldocs.com/bo/history.htm

Sears, Marian V., *Mining Stock Exchanges: An Historical Survey.* University of Montana Press, Missoula, 1973

Chapter Five

http://ceres.ca.gov/ceres/calweb/geology/goldrush.html

http://www.goldrush.com/~joann/women.htm

http://www.museumca.org/goldrush/fever15.html

http://www.museumca.org/goldrush/fever18.html
http://www.museumca.org/goldrush/fever19.html
http://www.museumca.org/goldrush/guest.html
http://www.ncgold.com/History/BecomingCA_Archive30.html
http://www.pbs.org/goldrush/changes.html
http://www.pbs.org/goldrush/discovery.html
http://www.pbs.org/goldrush/fever.html
http://www.pbs.org/goldrush/goldcountry.html
http://www.pbs.org/goldrush/journey.html
http://www.pbs.org/goldrush/sanfran.html
http://www.sfmuseum.org/hist2/gold.html

Chapter Six

http://www2.ccnmatthews.com/scripts/ccn-release.pl?/current/1111042n.html
http://www.currentcapital.com/7994.html,
http://www.mpm.edu/research/geology/gold_lore05.html
http://www.nap.edu/html/hardrock_fed_lands/appA.html
http://www.newmont.com/en/gold/howmined/index.asp
http://www.rhosybolbach.freeserve.co.uk/modrnmining.htm
http://www.simplifygoldstocks.com/Questions/FAQ.php
http://www.thebullandbear.com/articles/1998/899-orex.html
http://www.treasurefish.com/gold-outlook.htm
http://www.usgs.gov

Chapter Seven

Corti, Christopher, and Holliday, Richard, "A Golden Future," *Materials World*, February 2003, pp. 12-14.

Dines, James, *Invisible Crash* (special report), 2003

Dobra, Dr. John L., *The U.S. Gold Industry 2000*, Mackay School of Mines, University of Nevada (Reno). (See www.goldinstitute.org, Production and Jobs).

Doran, Ursel S., "All Gold In the Ground Is Not Created Equal: A Review of Gold Industry Valuation Fundamentals." (See http://www.gold-eagle.com/editorials_00/doran052900.html)

Lucas, John M., *Gold*, 1994. See http://minerals.usgs.gov/minerals/pubs/commodity/gold/3000494.pdf.

Olmsted, John A. and Williams, Gregory M., *Chemistry: A Molecular Science*, 3rd ed, John Wiley & Sons; 2001, p. 935-936.

Starchild, Adam, *Portable Wealth: The Complete Guide to Precious Metals Investing*, 1998, Paladin Press; January 1998, p. 4.

Chapter Eight

See endnotes below. Other references:

Bernstein, Peter L. , *The Power of Gold, The History of Obsession,* (New York, NY: John Wiley & Sons, 2000)

Green, Timothy, " Central Bank Gold Reserves: An Historical Perspective since 1845," World Gold Council Website

World Gold Council, "Background To Gold As A Reserve," World Gold Council Website

World Gold Council, "Gold As A Reserve Asset," World Gold Council Website

Mayer, Martin, *The Fed,* (New York, NY: The Free Press, 2001)

Weisweiller, Rudi, *How the Foreign Exchange Works,* (Paramus, NJ: New York Institute of Finance, 1990)

Salsman, Richard M., *Gold and Liberty,* (Great Barrington, MA: American Institute for Economic Research, 1995)

Chapter Nine

See endnotes below.

Chapter Ten

Barron's, *Terrorists and Texas Oil,* Oct. 15, 2001 http://interactive.wsj.com/archive/retrieve.cgi?id=SB1002927557434087960.djm&template=barrons.tmpl

Bernsek, Anna, *The $44 Trillion Abyss*, Fortune Magazine, Nov. 24, 2003, pages 113-116

Birnbaum, Jeffrey H., *$10 Trillion in Deficits*, Fortune, 9/10/2003

Brown, Ken, *Some Fear Inflation Is Ready for a Comeback*, WSJ, 12/1/2003, Figures 9, 10, 13, 14A, B

Business Week (Rich Miller, Michael Arndt, Kerry Capell, David Fairlamb), *The Incredible Falling Dollar*, 12/22/2003

Epstein, Gene, *Don't Give Up on the Dollar*, Barron's, 9/29/2003

Fleming, Mali, *Gold Rush*, WSJ, 12/2/2003

Grant, James, *The Money Printers*, Forbes, 6/9/03

http://interactive.wsj.com/archive/retrieve.cgi?id=SB906778372881209000.djm&template=barrons.tmpl

http://online.wsj.com/barrons/article/0,,SB1042850639116463264,00.html

http://online.wsj.com/documents/indicate.htm

http://prudentbear.com/archive_comm_article.asp?category=Credit+Bubble+Bulletin&content_idx=23958

http://www.berkshirehathaway.com/2002ar/2002ar.pdf

http://www.federalreserve.gov/boarddocs/hh/2003/february/ReportSection1.htm

http://www.swiftenergy.com/SFY/Investor-Info/Industry-Outlook/2001/HistO&G/AR99OGme.htm

Mayer, Martin, *The Fed*, The Free Press NY, NY, 2001

New York Times 2003 World Almanac, Edited by John W. Wright

Nolan, David, *Contemplating the Evolution from the Way We Were to the Way It Is*, Prudent Bear.com

Sicilia, David B., and Cruikshank, Jeffrey L., *The Greenspan Effect*, McGraw Hill, 2000

Uchitelle, Louis, *Why Americans Must Keep Spending*, New York Times, 12/1/2003

WSJ, 1994 Year End Review

WSJ, *Foreign Cash Flow Is Vital To US—But Will It Last?*, 1/15/2004

WSJ (Randall Smith, Steve Swartz, George Anders), *What Really Ignited the Market's Collapse After Its Long Climb*, 12/16/1987, page 1,20

WSJ, *Tomorrow's Elderly Fuel Health-Care Spending and Strain the System*, 12/19/2003, Bernard Wysocki Jr.

Chapter Eleven

Kosares, Michael J., *ABCs of Gold Investing*, Addicus Books, Omaha Nebraska, 1997

Chapter Twelve

Calandra, Thom, *As Gold Brightens, Paper Version Lags*, CBS MarketWatch, 08/08/03. http://cbs.marketwatch.com/news/story.asp?guid=%7B087CEDF8%2DB3E4%2D4E67%2DB662%2D3A343A584353%7D&siteid=bigcharts

Calandra, Thom, *New Gold Fund on Track*, CBS Market Watch, 9/29/03. http://cbs.marketwatch.com/news/story.asp?guid=%7B14CCEF9E%2DC555%2D4B60%2D91D9%2D7A19D5C1F907%7D&siteid=bigcharts

http://www.mcwatters.com/english/Txt-09.html

http://news.morningstar.com/doc/news/0,2,94254,00.html

http://www.morningstar.com/

Larsenova, Marketa, *This ETF Doesn't Have Midas Touch*, Morningstar, 7/21/03

Smith, Andy, *It's the Mother of All Bullion Products*, Zeal Intelligence, May 2003. http://Zealllc.com

Endnotes

[1] Bernstein, Peter L., *The Power of Gold: the history of an obsession*, 2000, New York: John Wiley & Sons, Inc., 121.

[2] Bernstein 113.

[3] Andrews, Kenneth R., *The Spanish Caribbean: Trade and plunder 1530-1630*, 1978, London: Yale University Press Ltd., 6.

[4] Díaz, Bernal, *The Conquest of New Spain* (translated by J.M. Cohen), 1963, London: Penguin Books Ltd., 104.

[5] Díaz 141.

[6] Díaz 148-149.

[7] de Sahagún, Fr. Bernardino, *The Florentine Codex*, as quoted by Marshall C. Eakin, Ph.D. (Associate Professor of History at Vanderbilt University), in *Conquest of the Americas* (a 24-lecture series published by The Teaching Company), lecture eight.

[8] Díaz 229.

[9] Díaz 274.

[10] Díaz, Bernal, as quoted by Eakin, lecture eight.

[11] Bernstein 123.

[12] Kamen, Henry, *Empire: How Spain became a world power 1492-1763*, 2003, New York: HarperCollins Publishers Inc, 107.

[13] Kamen 108-109.

[14] Kamen 117 (although he mistakenly quotes Poma as describing Columbus: Poma actually mentioned Pizarro by name).

[15] Kamen 501.

[16] Williams, Neville, *The Sea Dogs: Privateers, Plunder and Piracy in the Elizabethan Age*, 1975, New York: Macmillan Publishing Co., Inc., 57.

[17] Kamen 88.

[18] Williams 23-25.

[19] Williams 41.

[20] Andrews, Kenneth R., *Drake's Voyages: A re-assessment of their place in Elizabethan maritime expansion*, 1967, New York: Charles Scribner's Sons, 45-46.

[21] Andrews, *Drake's Voyages*, 104.

22 Martin, Colin, and Parker, Geoffrey, *The Spanish Armada*, 1988, New York: W.W. Norton & Co., 274.

23 Martin and Parker, 272.

24 Martin and Parker, 236.

25 Martin and Parker, 259.

26 Martin and Parker, 257.

27 Martin and Parker, 258.

28 Martin and Parker, 260.

29 Martin and Parker, 135.

30 Williams 206.

31 Bernstein 135, emphasis added.

32 Kamen 474.

33 Bernstein, 331

34 Bernstein, 333

35 Bernstein, 334

36 Bernstein, 346

37 Bernstein, 340

38 Bernstein, 340

39 Bernstein, 329

40 Bernstein, 329

41 Bernstein, 339

42 Bernstein, 339

43 Bernstein, 330

44 Ware, Dick, *Research Study 26: The IMF and Gold (Revised Edition)*, the World Gold Council, 19

45 Quoted in Bernstein, 353

46 Bernstein, 355

47 Bernstein, 357

48 Bernstein, 358

49 Bernstein, 363

50 Bernstein, 363-364

51 Kirk, Donald, *Korean Crisis: Unraveling of the Miracle in the IMF Era*, 1999, New York: St. Martin's Press, 39-40

52 Blustein, Paul, *The Chastening: Inside the Crisis that Rocked the Global Financial System and Humbled the IMF*, 2001, New York: PublicAffairs, 338. At the time, Brazil had a GDP of $800 billion.

53 According to an advertisement run by the World Gold Council, run in the *New York Times*, September 8, 1998. Quoted in Bernstein, 364.

54 Blustein, 20-21

55 Ware, 42

[56] Ware, 37
[57] Kirk, 17
[58] Bernstein, 366
[59] Ware, 36
[60] Bernstein, 372

Index

3D modeling 88
49ers 62, 63

A

Aguilar, Jeronimo de 15
Alexander the Great 5, 6, 13, 173
Alvarado, Pedro de 15, 19
Alzola, Captain Tomás de 13, 27
AngloGold LMT 78
Aristotle xi
Armada, Spanish 27–30
Armour, Philip 64
Asian contagion 114, 125
assignat, French 48
Atahualpa 20, 21
Aztec Empire, conquest of 15–20, 22

B

Babylonia 4
Bank of England 11, 12, 96, 103, 115
Barrett, Robert 26
Barrick Gold Corporation 77, 78, 79, 84, 167
Belgium, losses to Nazis 38
bezant 7, 8, 12
Bible 3
bills of exchange 10, 11
Blackbeard 33
Black Bart 33
Boer War 51–52
Brannan, Sam 59, 64
Bretton Woods 56, 97, 98, 101, 114
buccaneers 32, 33
Buffett, Warren 122, 123, 174
Byzantium 7, 8

C

Captain Kidd 33
Cavendish, Thomas 26, 27
central banks
  1999 moratorium on gold sales 95
  gold reserves 104–105
  Joint Statement on Gold 103
  Nazi thefts from European 37
  origins of 96
  reasons for holding gold 106–107
Charles V 16, 20
China 7, 30, 44, 55, 62, 82, 109, 115, 145, 175, 176
civilians, Nazi thefts from 38
Civil War, U.S. 51
closed-end funds 164–165
coin, world's most expensive 154
coin clipping 10
Colchis 4
Columbus, Christopher 13, 14, 15, 23, 80, 188
confiscation of gold in U.S. 55, 151
contributors vii
Cortés, Hernán 16, 17, 18, 19, 20, 23
Crassus, Marcus 13
Croesus 4, 5, 12

D

Darius 5
debasement, coin 6, 7, 10, 173
deficit 10, 106, 117, 126, 127, 128, 130, 136, 137, 141, 174, 175, xii

denarius 6
derivatives 122, 122–124, 174
Dias, Bartolomeu 14
Díaz, Bernal 15, 17, 18, 19, 188
Dionysius 6, 173
dollar
  and the failure of U.S. fiscal discipline 98
  dependence on gold for strength 98
  fall in index, 1986-2004 ix
  implications of falling 174
  made centerpiece of world financial system 97
  massive debasement of 98, 99
  reasons for continued decline of 135–136
  strong-dollar policy now history 114
Dow industrials 141
Drake, Sir Francis 25, 26, 27, 30, 33, 188
ducats 8, 9, 27
Dutch West India Company 31

E

EGTF (Equity Gold Trust Fund) 165, 176
Egypt 3, 5, 80, 110, 155, 173
Eighty Years' War 31, 34
Elizabeth, Queen 24, 26
Enriquez, Don Martin 25, 26
epithermal modeling 88
Equity Gold Trust Fund. *See* EGTF
euro 103

F

Farouk, King 155
Federal Reserve, U.S. 55, 102, 117, 121, 122, 124, 125, 126, 130, 134, 137, 174, 175, xii
  changing role of 121–122
  losing power to manipulate economy 174, 175
Fenton, Stephen 155
Ferdinand, King 14, 15
Flanders, Army of 27–28
Florence 8, 9
florin 9, 12
Franks 7

G

gangue 72, 73
de Gaulle, Charles 100, 115
Genoa 8
genoin 8, 12
gold
  as the perfect money xi
  attributes and uses of 80–82
  central banks, reserves of 104–105
  colloidal 82
  confiscation of. *See* confiscation of gold in U.S.
  demand for 83–84
  direct investing in 142
    American Eagle 145
    Australian Kangaroo 145
    Canadian Maple Leaf 145
    Chinese Panda 145
    gold bars 143–144
    how to build a U.S. gold set 149–156
    how to buy coins 146
    how to buy rare U.S. coins 149
    private mint coins 144
    South African Kruggerand 144
  exploration
    techniques 87–88
      3D modeling 88
      epithermal modeling 88
      infrared mineral analysis 88
      litheogeochemistry 88
      paleontology 88
    undiscovered gold worldwide 89–94
  how much to buy 177–178
  indirect investing in
    closed-end funds. *See* closed-end funds
    EGTF (Equity Gold Trust Fund). *See* EGTF

gold stocks. *See* stocks, gold
mutual funds. *See* mutual funds, gold
influence of private owners on world markets 114
key role in IMF 112–113
largest nuggets found x
medium of exchange, as 79, 83, x
prices, long-term 176
price predictions 141–142
production
advances in mining 89
economics of 86–87
grades of deposit 85, 86
Haber Gold Process (HGP) 89
historic data 74–76
majors vs. juniors 78
modern world 76–78
North American 76–77
South African 76, 77
top 5 mines in world 77
total in last 6,000 years 85
units of 74
variables affecting 74
properties of xi
reasons for central banks to hold 106
reasons for historical value 117, x–xi
recovery of
flotation 72–73
leaching 73–74
recycling of 80–81
removal from world monetary system 97
rise in price of, 1972-1980 101
rise in price of, 1999-2004 ix
Golden Fleece 4, 173
gold pool 99, 115
gold standard 47, 48, 49, 50, 52, 54, 55, 56, 57, 96, 100, 106, 115
U.S. abandonment of 55, 100
Greenspan, Alan 107, 135, 175, 187
guinea 11, 12, 47
Gyges 4

H

Hawkins, Sir John 23, 24, 25, 26
Hebrews 4
Heyn, Piet 31
HGP (Haber Gold Process) 89
Hispaniola 14, 15, 23, 32
Hitler, Adolph 35, 36, 37, 38, 39, 40, 41, 43, 45, 183
Homestake 79
Hoover, President Herbert 98
Huascar 20, 21
Huguenots, French 31
hyperinflation
definition of 52
in last quarter of 20th century 56
post-revolution: French 48
post-WWI: other nations 53
post-WWII 55–56

I

IMF (International Monetary Fund) 101, 102, 104, 105, 107, 109, 110, 111, 112, 113, 114, 115, 116, 189
bailouts of troubled nations 109, 111
gold's role in 112–113
gold sales of 102, 109, 115, 116
Incan Empire
conquest of 20–22
treasure of 21
India 5, 6, 14, 31, 47, 115, 176
inflation 6, 7, 10, 51, 52, 54, 56, 102, 103, 124, 127, 130, 131, 133, 134, 135, 136, 137, 141, 166, 174, 175, 184, xii
infrared mineral analysis 88
interest rates 103, 121, 124, 127, 130, 134, 136, 137, 166, 174, 175
International Monetary Fund. *See* IMF
Isabella, Queen 15
Italy
losses to Nazis 38
rise of merchants 8

J

Japan, theft of gold in WWII 44–45
Johnson, President Lyndon 99

K

Keynes, John Maynard 96, 97, ix
Klondike 67

L

lithogeochemistry 88
Lombards 7
LTCM (Long Term Capital Management) 125, 126
Lydia 4, 5, 173

M

Madre de Dios, capture of 30
mandat, French 48
Marcos, Ferdinand 45
Medina, Duke of 29, 30
Merkers, Nazi gold in 35, 41, 42, 43, 44, 45, 183
Merrill Crow unit 73, 74
mining, hydraulic 66
moneyers 8
Montezuma 16, 17, 18, 19
Morgan, Henry 32, 33
Morgenthau, Henry Jr. 96
Muslims 7, 8
mutual funds, gold 157–165
  benefits of 158–160
  how to select 160–162
  list of 163–164
  ways to profit from 157

N

Narvaez, Panfilo 19
Nero 7
Netherlands, losses to Nazis 38
Newmont Mining Corporation 77, 78, 167, 169
Newton, Sir Isaac 47, 48
NGC (Numismatic Guarantee Corporation) 149, 150, 151, 152, 153, 154, 155
Nixon, President Richard 100, 101
Norway, rescue of gold from Nazis 45

O

oil 118–121
  dependence on 173
  producers of and the dollar 175

P

paleontology 88
palladium 80
Parma, Duke of 28
Parthia 13
PCGS (Professional Coin Grading Service) 149, 150, 151, 152, 153, 154, 155
Persia 5, 173
Philip II, King 24, 25, 27, 28–31, 29, 30
Pindar ix
pirates 13, 26, 27, 32, 33
Pizarro, Francisco 20, 21, 22, 23, 188
platinum 43, 80, 81, 91, 159, 168
Polk, President: confirmation of California gold 61
Polo, Marco 14
Poma, Guaman 22, 188
pound sterling 6
Pratt, Bela Lyon 151
privateers 25, 26, 27, 30, 31
proof coins 155

R

recessions
  and oil 118, 173
Reichsbank 35, 38, 39, 41, 42
Reneger, Robert 23
reparations, return of Nazi gold as 39
Rome 6, 7, 155
Roosevelt, President Franklin 43, 55, 151
Roosevelt, President Theodore 150, 151, 153

S

S&P
fall in, 2000-2002 ix
Saint-Gaudens, Augustus 153
Saint Gaudens 147, 148, 149, 154, 155, ix
Santa Ana, capture of 13, 27
San Francisco 59, 60, 63, 64, 150, 152, 153, 154
San Juan de Ulua, battle of 25–26
San Salvador, ambush of 23
Saudi Arabia 118, 119, 120, 173
shekel 4
shilling 6, 11
silver 3, 4, 5, 6, 8, 9, 10, 11, 13, 22, 23, 26, 29, 30, 31, 33, 42, 43, 47, 48, 55, 80, 81, 82, 91, 150, 159, 164, 167, 168
solidus 6, 7
stater 4, 5, 6, 12
stocks, gold 165–170
stock market 106, 114, 122, 123, 124, 134, 136, 160, 174, 175, 177, ix
Strauss, Levi 64
Studebaker 64
sumptuary laws 9
Sutter, John 59, 60, 61, 63
Switzerland
bank accounts of Nazi victims in 39–40
complicity with Nazis 40–41
reparations to victims of Nazis 41, 44

T

Tenochtitlan 16, 17, 18, 19, 20
TGC (Tripartite Commission for the Restitution of Monetary Gold) 44
Tlascalans 17
tonne, definition of 74
trading partners, Nazi 40
Turk, James 141
Tut, King 3, 80, x

U

U.S. Geological Survey (USGS) 74, 88, 90, 91
United States
1994 meltdown in Treasury bonds 124
aggregate debt 130–131
ballooning debt and deficits 126–130, 175
current economic problems 117
growing inflation in 131–135, 136–137, 141–142
reasons for continued decline of dollar 135–136
squandering of national wealth 99
Treasury sales of gold 101, 115

V

Velázquez, Diego 16, 19
Vietnam War 99
Volcker, Paul 102

W

Wells Fargo 64
Wheaton River Minerals 79
White, Harry Dexter 97
women, opportunities for during gold rush 65
World Gold Council 80, 83, 84, 87, 90, 104, 112, 116, 165, 186, 189
World Jewish Congress 40, 44
World War I, monetary effects of 52

Y

Yukon 67

Edge Financial Company, Inc.
9730 SW Cascade Blvd., #200
Tigard, OR 97223-4324